Roofing
the Right Way
2nd Edition

Roofing
the Right Way
2nd Edition

Steven Bolt

Drawings by Gregory W. Thorne

TAB Books
Division of McGraw-Hill, Inc.
Blue Ridge Summit, PA 17294-0850

SECOND EDITION
FIFTH PRINTING

© 1990 by **TAB Books**
First edition © 1986 by TAB Books
TAB Books is a division of McGraw-Hill.

Library of Congress Cataloging-in-Publication Data

Bolt, Steven, 1949-
    Roofing the right way / by Steven Bolt.—2nd ed.
        p.   cm.
    ISBN 0-8306-7387-3   ISBN 0-8306-3387-1 (pbk.)
    1. Roofing. I. Title.
TH2431.B58 1990
695—dc20   89-49451
                                                    CIP

Acquisitions Editor: Kimberly Tabor
Book Editor: Joanne M. Slike
Direction of Production: Katherine G. Brown
Book Design: Jaclyn J. Boone   HT3-4
Manufactured by Arcata Graphics-Fairfield, Fairfield, Pa.   3387

For Shirley and Julie
and those who remember them

Contents

Acknowledgments

*T*his book is equally a product of my roofing experience and my interest in words. While it might appear that roofing and writing are dissimilar skills, I am fortunate to have colleagues and friends who have shown me that both disciplines require precision and craftsmanship.

I thank Ken Bolt, Daniel Clem, Chet and Julia Dawson, Tom and Anna Farnkoff, Douglas Hartz, Ed McCarthy, Marilyn McDonnell, Kip Naughton, Roland Phelps, Sporik Roofing, R. F. Fager Company, and the employees of TAB Books. I wish to thank Onduline, Inc. and Patricia Pitts for providing the illustrations that appear in chapter 11.

Thanks also go to Carolyn Anderson, Jackie Boone, Suzanne Cheatle, Jean Fiori, Susan Hansford, Patsy Harne, Richard Hawkins, Rita Henderson, Linda King, Lisa Mellott, Nadine McFarland, Douglas Robson, Charles Sanders, Kimberly Shockey, Joanne Slike, Kimberly Tabor, Greg Thorne, Leslie Wenger, Joan Wieland, and Bob Ziegler for lending their talent and patience. Special recognition is due Ray Collins for providing the title alliteration.

The first edition of this book is dedicated to Deborah Shaw, with special thanks to John J. Romano. *Amícus certus in ré incertá cenitur.*

About This Book

*R*oofing the Right Way describes the materials, tools, and step-by-step application techniques for successfully installing roofing materials on a new building, over one layer of worn shingles, or where worn shingles first must be torn from the roof deck. Beginning with guidelines on how to determine if your roof needs to be replaced, this book presents details on the advantages of using specialized roofer's tools and equipment, the most common roofing materials—asphalt fiberglass-based shingles, wood shingles, wood shakes, metal roofing, and roll roofing—and how to waterproof roof-line intrusions and obstacles such as chimneys, walls, valleys, and vents.

Because about 70 percent of all homes are ventilated inadequately and because the installation of roof ventilation devices from turbines to skylights is easiest during reroofing, I have included instructions for installing a balanced home-ventilation system.

If installing a new roof or reroofing over worn shingles at first seems to be an intimidating task, consider that you can save as much as two-thirds the cost of keeping your home waterproof by doing the shingling yourself. If you are uncertain about taking on your own roofing project, the information presented here will help you understand the nature as well as the details of the work required. With this book, at the least, you will be able to discuss your roofing needs with suppliers and prospective contractors. Technical terms mentioned throughout the book are defined on first usage and compiled in the Glossary. Addresses of equipment manufacturers and product suppliers named in the text are listed in the Sources appendix.

Glance through the chapters that describe the application of the roofing materials you want to use on your home. You will find that the work is well within the capabilities of do-it-yourselfers. Within these pages, you will also find "Application Tip" subheadings. These practical tips provide advice that is based on years of professional experience. From the examples explaining how to estimate materials to the definitions of technical terms, the emphasis is on carefully written and technically precise descriptions.

Chapter 1

Conditions and Materials

According to United States Census Bureau statistics, about 23 million single-family homes are at least 50 years old, and several more million are at least 15 years old. With approximately 80 percent of American homes roofed with asphalt-based shingles or asphalt fiberglass-based shingles that are designed to last a minimum of 15 to 20 years, age alone is a primary reason that many homeowners must reroof.

Until there are roof leaks, caused by storms or aging shingles, most homeowners will not be prompted, or find it necessary, to make more than periodic cursory inspections of their roofs. Roof leaks often do result directly from storm damage or from loose flashing around a chimney or vent, but missing shingle tabs, or buckled, warped, or aged shingles also are indications that roof materials need to be repaired or replaced.

If your shingled roof is less than 15 years old and leaks are apparent, minor repairs probably will solve the problem. If the shingles on your home were installed 15 or more years ago, inspect the condition of your roof very carefully.

INSPECTION PROCEDURES

The cause of a roof leak is not always easy to find. It is not unusual for water to seep through cracked or worn flashing along a wall, chimney, or valley, and to show up on interior walls as water stains many feet from where the water first entered. With periodic, cursory inspections, you can usually spot signs of roof trouble. In many cases, you can use binoculars to make inspections. Binoculars are especially useful for inspecting potential problems with white or gray shingles, but dark-colored shingles or black shingles often will not reveal problems when viewed from the ground. If a problem is apparent because of water damage, be prepared to climb on the roof to determine the extent of the trouble and to inspect the shingles closely for less obvious, but just as serious, signs of worn materials.

To determine whether your roof will need repairs or a complete reroofing, inspect the roof surface with the following points in mind. If you find that your roof fits the descriptions in either or both of the last two categories, your home definitely needs new shingles.

- Look for missing tabs or shingles. A few shingles or tabs blown off during a storm usually can be replaced easily, but if a roof section is missing many tabs or entire shingles, and the damage is not from a recent storm, you probably have a worn-out roof.

- Check around skylights, vents, pipes, and valleys for cracked or worn caulking, roof cement, and metal. Intrusions in roofs are often the most frequently weather-stressed areas and, therefore, the most common locations where leaks originate. If the leak is not serious, such problem areas usually are not difficult to repair with roof cement.

- Closely inspect the flashing around chimney walls, valleys, and dormers. Look for cracked, worn metal and deteriorated roofing cement and loose nails or loose step flashing. Leaks around flashing are sometimes difficult to locate and fix.

- Check for green "mold" or "mildew" on shaded areas of your shingled roof. If you discover any, you could remove the fungus by spraying the area with a mixture of detergent and pressurized water. The cure, however, might prove more of a drawback than the benefit of an improved appearance for your roof. The fungus itself will not damage asphalt fiberglass-based shingles. You could dislodge a significant amount of surface granules and, therefore, lessen the expected life span of your shingles. The best way to prevent recurring fungi is to clear the shrubbery shading your home.

- Inspect the general condition of the shingles. Worn shingles will have tabs that are cracked or curled at the edges (FIG. 4-1). A significant loss of color and granules will be apparent. Large amounts of dislodged stony surface granules will appear in gutters and at downspout outlets where they have been washed off the shingles. When this top coat of granules wears off, patches of asphalt will begin to appear on individual shingle tabs. If pressure is applied to a piece of a shingle, it will easily break and crumble in your hand.

- Go to the roof surface and look quite closely at the granules where the patterned cutouts of the shingles fall. With shingles that are sufficiently aged, you will be able to see the wooden deck through very small holes that have worn through the shingles. Check several rows of the cutouts. If there are many similarly worn shingles, it is a sure sign that your roof needs to be reshingled.

SHINGLE LIFE SPAN

Shingles are manufactured in a wide variety of colors, styles, and weights. Because of the ease of application, three-tab, asphalt fiberglass-based

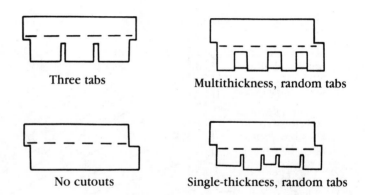

Three tabs　　　　　　Multithickness, random tabs

No cutouts　　　　　　Single-thickness, random tabs

1-1　The most common types of asphalt-based and fiberglass-based shingles.

shingles (FIG. 1-1) are by far the most common roofing material used on recently constructed or reroofed American homes.

Most manufacturers of asphalt fiberglass-based shingles offer shingles at 225, 260, 300, and 325 pounds per square. As the weight per square increases, so does the shingle life span and, not coincidentally, the purchase price. The lightest shingles are designed to last 20 years; the middle-weight range of shingles are designed to last 25 to 30 years; and the heaviest shingles are made to weather over 30 years.

Many local building suppliers will be familiar with and will stock only the most popular shingle weights and styles. Be prepared to shop around. Also be prepared for some of your questions to go unanswered.

If you live in a suburban development or an area where all the houses were constructed within a year or so of each other, and if quite a few of your neighbors have been reroofing their homes, you can reasonably deduce that your roof is ready for replacement. Nevertheless, don't be rushed into a hasty decision by the actions of neighbors. Inspect your roof carefully before deciding to make repairs or to completely reroof your home.

Salesmen for roofing contractors will often work an entire neighborhood and attempt to convince potential customers that every roof in the area is worn out. Some salesmen will strongly recommend that all old shingles on any home must always be torn off before a new layer of shingles is installed. Actually, the old shingles must be torn off only when they are buckled or warped or if the roof already has two layers of shingles. Whether you do the work or have a contractor do the job, tearing off shingles will greatly increase the time, effort, and cost required to complete the job.

Roof Pitch

The pitch of a roof (FIG. 1-2) is very much a factor in the life span of shingles. Generally, the steeper the angle of the roof, the longer the shingles

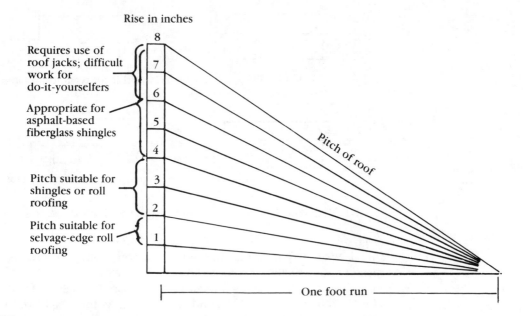

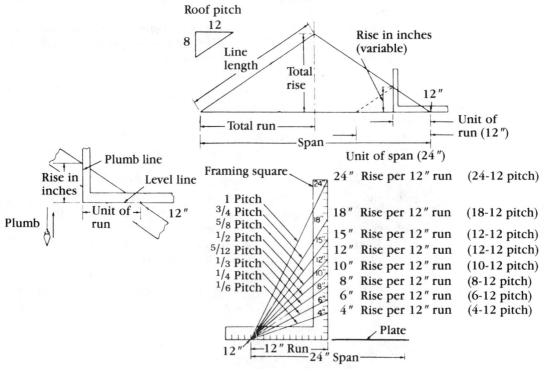

1-2 The pitch of a roof indicates the degree of the roof slant. Stated as a fraction, the pitch is a ratio of the rise to the span of a roof. Most homes have roofs that rise between 2 and 5 inches per foot.

will last. Suppose that you own a Cape Cod style of house and your next-door neighbor has a ranch house. If the same brand and weight of shingles are applied in the proper manner to both houses, the shingles on the ranch house will wear out several years before the shingles on the Cape Cod house. Because of the steeper pitch of the roof surface, the shingles will be exposed to less direct sunlight each day. In addition, rain running off the steeper slopes will generate less drag, and therefore, fewer granules will be dislodged.

The angle and intensity of the wind, rain, and especially sunlight as they strike a roof differ with the positions of buildings. The amount of sunlight and wind, and the frequency of freeze/thaw cycles a roof is exposed to relates directly to the number of years an asphalt fiberglass-based shingle roof will last. Even the shade from trees or the wind deflection provided by a hill can make a difference (FIG. 1-3). It is not unusual for the "sunside" of a roof to wear out several years before the "backside" of the same roof.

1-3   The wind deflection and the shade created by trees can add years to the life span of roofing materials. The roofing on the sun side of a house positioned like this will likely wear out several years before the shaded side.

Shingle Weight

One of the most important ways shingles are categorized is by weight per 100 square feet. Roofers commonly refer to 100 square feet of roof coverage as a *square*. Until the introduction of shingles made with fiberglass, it was accurate to presume that greater weight in a shingle meant longer life for the shingle. However, the combination of fiberglass and asphalt in shingles now permits lighter weight to be combined with longer shingle life.

The lightest, three-tab, asphalt fiberglass-based shingles weigh 225 pounds per 100 square feet, have a 20-year limited warranty, and are classified by Underwriters Laboratories as class A. Class A is the highest-possible fire-resistance rating for shingles.

The lightest, three-tab, asphalt shingles weigh 235 pounds per 100 square feet. They have a 15-year limited warranty and are classified by Underwriters Laboratories as class C. Class C is the fire-resistance rating for most conventional asphalt shingles.

Table 1-1 Styles, Weights, and Dimensions of Roofing Materials.

Material	Configuration	Per Square			Size		Exposure in inches	Specifications
		Weight in pounds	Shingles	Bundles	Width in inches	Length in inches		

Fiberglass shingles

| 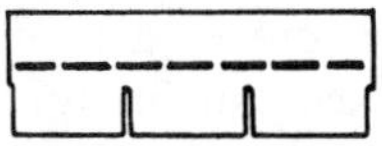 | 3 tabs | 210 or 215 | 78 or 80 | 3 | 12 or 12¼ | 36 | 5 or 5⅛ | UL Class A |

Organic, three-tab shingles

| 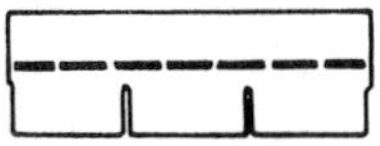 | 3 tabs | 235 to 290 | 78 or 80 | 3 or 4 | 12 or 12¼ | 36 | 5 or 5⅛ | UL Class C |

No-cutout shingles

| 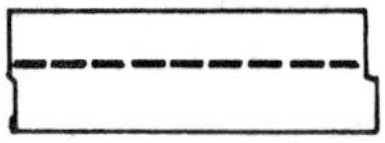 | various | 235 or 240 | 78 | 3 | 12¼ | 36 | 5 | UL 55B |

| 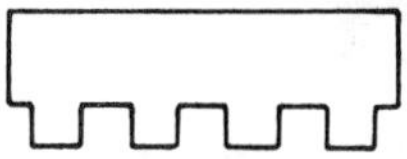 | tabs give wood-shake look | 250 or 300 | 100 | 4 | 12 | 36 | 4 | UL 55B |

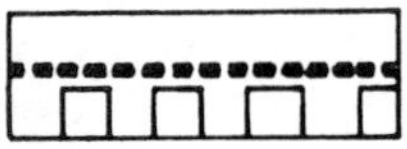	two-ply strip shingles give layered look	AMERICAN STANDARD						UL Class A
		320	84	4 or 5	11½	36	4¾	
		METRIC						
		320	66	4 or 5	13¼	39⅜	5⅝	

ASPHALT AND FIBERGLASS

Produced in many colors, textures, and weights, the three-tab-strip, asphalt fiberglass-based shingle is the type of roofing material most commonly available through local suppliers throughout the country. Some manufacturers offer asphalt fiberglass-based strip shingles designed with a layered texture that resembles wood shakes, shingles with two cutouts, or strip shingles with no cutouts. (See TABLE 1-1 for a comparison of styles, weights, and dimensions.) Although asphalt fiberglass-based shingles are

Table 1-1 Continued.

Material	Weight		Square feet per package	Length in feet	Width in inches	End lap in inches	Top	Exposure	Specifications
	Per roll	Per square							
Mineral-surface roll roofing	75 to 90	75 to 90	About one	36 or 38	36	6	2 or 4	34 or 32	
Double coverage, mineral-surface roll roofing	55 to 70	55 to 70	one-half	36	36	6	19	17	
Coated roll roofing	43 to 65	43 to 65	one	36	36	6	2	34	43, 55, or 57 pound
									65 pound
Felt	52	15	4	144	36				15 pound
						4 to 6	2	34	
	60	30	2	72	36				30 pound

susceptible to cracking if they are installed during cold weather, the light weight of the fiberglass allows for a lighter shingle coupled with a longer limited-warranty period. The fiberglass also allows for greater fire resistance than that provided by asphalt shingles.

Asphalt shingles, or so-called *composition shingles,* are made with a layer of heavy roofing felt (made from organic materials such as paper or wood chips) that is saturated with asphalt. A thick coat of asphalt and minerals is then added to the saturated felt. The next step is to add a layer of ceramic granules or opaque-rock mineral granules for color and for weather and sunlight resistance. In addition, strips of adhesive are added to the face of each shingle. After the shingles have been properly installed on the roof, the heat of the sun activates the adhesive to seal down each shingle.

Few shingle manufacturers now produce asphalt-based shingles. They have been supplanted in the market by asphalt fiberglass-based shingles. Asphalt fiberglass-based shingles are manufactured using a process quite similar to that for asphalt shingles. Essentially, the difference is the use of an inorganic fiberglass base that is saturated with asphalt. Fiberglass-based shingles also feature self-sealing strips of adhesive.

A wide variety of shingle colors is available. White, off-white, light gray, or very light pastel colors on your roof will reflect more sunlight during the summer than black or dark shades of asphalt-fiberglass shingles. Conversely, the light colors will absorb less sunlight during winter than dark shingles. If either summer cooling or winter heating is an unusually important factor at your location, select an appropriate color. Otherwise, pick a color that suits the scheme for the house.

Chapter **2**

Tools and Equipment

Having the proper tools and equipment is a very important part of any roofing work. Some of the specialized items you will need might be somewhat difficult to locate. Even well-stocked hardware stores don't carry all types of roofing tools. Plan to have all your tools before you actually start nailing shingles.

When the weather is hot, wear lightweight clothing such as a loose-fitting shirt and jeans. To keep your head cool and prevent sunburn, wear a sun visor or a cap. A sweatband for your forehead also will be useful. For cold-weather work, wear several layers of clothing. For increased dexterity, cut the fingers and thumb from the cotton glove for the hand you use to handle nails; leave the other glove intact.

HAND TOOLS

As a minimum for shingling even the easiest of roofs, you will need a roofer's hatchet or a carpenter's hammer, a nail pouch, a utility knife and blades, a chalk line and chalk, a tape measure, shoes with rubber soles, and a ladder.

Broom or Brush For tear-off work, you will need a common broom to sweep away debris, shingle granules, and nails from the exposed roof deck. A stiff-bristle brush can be used to spread roofing cement for roll roofing.

Carpenter's Hammer The claw on a carpenter's hammer (FIG. 2-1) is very useful for removing or loosening vent flanges and pulling out old, stubborn flashing and nails. The smooth surface of the head of a carpenter's hammer is designed for driving nails flat to the surface of wood without leaving a mar. *Claw hammers,* or *ripping hammers,* can be used for shingling. Nevertheless, a round, lightweight head is a disadvantage for driving roofing nails using the rapid, rhythmic technique favored by skilled roofers.

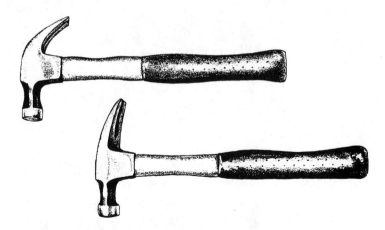

2-1   A claw hammer (top) or a ripping hammer is useful for prying flashing or nails. Either tool can be used to nail shingles, but a roofer's hatchet is more appropriate for such work.

Any do-it-yourself roofer who believes there is such a thing as a multipurpose hammer can expect bruised fingers and slow work. Don't use a tack hammer, a mallet, or a ball-peen hammer for shingling.

Caulking Gun   Use caulk (FIG. 2-2) to enhance waterproofing of flashing around a pipe flange or at a chimney where counterflashing is used. Do not use caulk as a substitute for roofing cement.

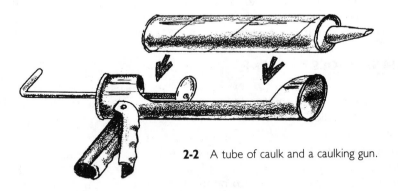

2-2   A tube of caulk and a caulking gun.

Chalk and Chalk Line   Blue, red, or yellow chalk is most often sold in plastic tube containers (FIG. 2-3). The chalk, the chalk box, and the chalk line must be kept dry. When you fill a chalk box, pull about half of the line out of the chalk box and pour in only enough chalk to fill the box halfway full. If you overfill the box with chalk, the line will jam. Rewind the line and be careful to avoid causing knots or tangles. Pull out the line about 2 to 3 feet to see if the chalk has saturated the line. If not, pull the line out halfway again and add a little more chalk.

2-3 A chalk line and a tube of chalk.

When you are snapping lines on a roof, it is generally possible to snap four or five lines before you must rewind the line. It will not be necessary to refill the box with chalk each time you rewind the line.

Gloves   Cotton work gloves (FIG. 2-4) are essential for tearing off worn shingles. You also will find such gloves useful when you unload shingles from the delivery truck.

2-4   For tear-off work, wear cotton work gloves. To maintain dexterity during cold weather, the glove fingers and the thumb can be cut from the nail-holding, gloved hand.

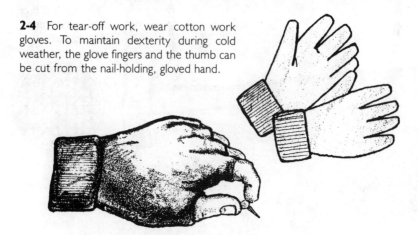

Nail Pouch or Apron   A nail pouch (FIG. 2-5) should be large enough to hold several handfuls of roofing nails. Making excessive trips to the nail box wastes time and energy. Larger pouches often have separate compartments that are convenient for holding a utility knife and a tape measure.

2-5 A nail pouch with a loop for holding a hatchet, small compartments for holding a rule, a utility knife, pencils, and a large compartment for nails.

Roofer's Hatchet   A roofer's hatchet is designed for professional roofers who typically drive a 1-inch roofing nail with one stroke. The square head is large and heavy and it has a gridded surface (FIG. 2-6). Some roofing hatchets have a sharp blade for splitting wood shakes. Other hatchets have a small knife blade for cutting asphalt-based shingles. It is much easier to cut yourself with the small, sharp blade than it is to use it to cut shingles.

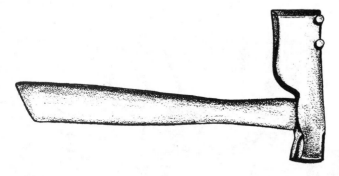

The milled face helps prevent flying nails

2-6   A shingling hatchet has a large, square head with a milled face. The adjustable gauges provide the roofer with a handy means of checking the alignment and exposure of a shingle as it is positioned and nailed.

Safety Eyeglasses   Many hardware stores carry inexpensive, plastic safety goggles, face shields, and other types of safety glasses. Some shields are designed to fit over the top of standard prescription eyewear. Regular prescription glasses and contact lenses do not provide protection from nails, staples, or pieces of wood or metal thrown up by a roofer's hatchet or by power tools. In the process of installing shingles on a typical roof, you can be certain that a number of roofing nails will come "shooting" at you as the inevitable result of striking some glancing hammer blows. Safety goggles that meet American National Standards Institute requirements are marked with a "Z87" on the eyewear.

Shoes Don't underestimate the importance of wearing rubber-soled shoes (FIG. 2-7) while you are working on your roof. Other types of shoes will not give you the proper footing you need to work on a sloped roof. Proper shoes are of utmost importance for working on steeper-sloped roofs. (See FIG. 1-2.)

This type of shoe will easily mar shingles   Wear soft-soled shoes

2-7 Don't overlook the importance of wearing the proper footwear. The wrong type of shoes (left) will mar shingles. Wear soft-soled shoes that won't slip.

Another factor is that hard-soled shoes will quickly mar new shingles. Check to see if your shoes have large, metal hooks for securing the laces. Use tin snips to cut off such hooks. If you do not remove the hooks, you will continually mar the shingles by dragging your feet over the shingles as you work in the proper roofer's sitting position.

Shovel A flat-bladed, tear-off shovel (FIGS. 2-8 and 4-4) is the only type of shovel to use when you must remove worn layers of shingles from a roof. Don't try to use a garden shovel to tear off shingles. Remember to wear gloves for tear-off work.

2-8 Use a flat-edged shovel for tearing off shingles.

Tape Measure   A tape measure (FIG. 2-9) doesn't have to be elaborate, but it does have to be accurate. A 50-foot tape is convenient for measuring the total square feet of a roof. A 6- or 8-foot tape will do the job just as well if you are very careful when you make measurements.

2-9   A lockable rule will be adequate for most roofing-work measurements. A 50-foot tape will be useful for measuring where roof-pattern tie-ins are required.

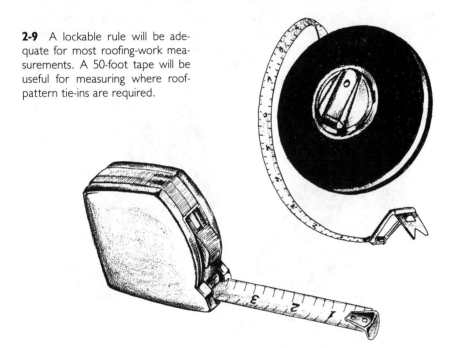

Tin Snips   In addition to cutting metal for flashing and valleys, so-called tin snips or aviation snips (FIG. 2-10) can be used to cut shingles. For the inexperienced roofer, cutting shingles with a utility knife is often a most difficult task to learn. To avoid scraped knuckles or worse injuries to your hands, snap a chalk line to obtain a straight line and cut the roofing material with tin snips. Ask your tool supplier for snips designed for either right-hand or left-hand use.

2-10   When you purchase tin snips, look for a pair designed specifically for a left-handed or a right-handed person. Both types are manufactured.

Trowel   A trowel (FIG. 2-11) makes easy work of applying roofing cement. Use the flat edge of the trowel to smooth out an even layer of roofing cement around a vent or chimney.

2-11   A trowel for applying roofing cement.

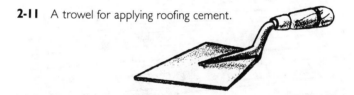

Utility Knife and Blades   The roofing tool that is the most difficult to learn to use properly and skillfully is the utility knife (FIG. 2-12). Straight blades—used primarily by carpenters—are easy to find in stores, but hook blades are the best type of utility blades to use for cutting shingles. Make sure that the blades are properly secured in the knife. Be careful!

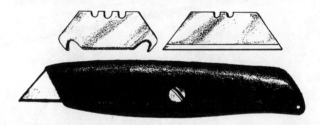

2-12   A utility knife, a hook blade, and a straight blade. Hook blades are better for cutting asphalt or fiberglass shingles.

ADDITIONAL TOOLS

For complicated roofing work, additional tools will be needed. Use the following descriptions for determining the tools and equipment you will need for the kind of building on which you will be working.

Roof Jack.   If the pitch of your roof is too steep for you to walk on comfortably, use roof jacks (FIGS. 2-13, 2-14, and 2-15) and planking as a safety precaution. Some roofs are so steep that roofing them without jacks is impossible. If your roof seems even a little too steep to walk comfortably, you might want to use jacks to hold the bundles of shingles after you bring the bundles up from the ground. Always use spikes (3-inch, 10d nails) to secure roof jacks. The jacks are designed so that they can be removed easily. When it is time to move the jacks and planking to a different part of the roof, it will be easier to drive the spikes into the deck instead of pulling them free.

Staple Guns   Pneumatic staple guns (FIG. 2-16), an air compressor, and hoses are much too expensive to purchase for limited use. Nevertheless, renting power tools might be worthwhile if your roofing job—such as multiunit apartment buildings—is large enough to justify using such

2-13   On a moderately pitched roof, jacks and planks can be used to hold a few bundles and the roofer.

2-14   On a steeply pitched roof surface, roof jacks and planks must be used.

2-15 This type of roof jack can be adjusted to accommodate the steepness of the roof pitch.

Powered
shingle stapler

2-16 A roofing stapler.

equipment. A staple gun will double the amount of shingles you can install in one day.

The Duo-Fast RS-1748 roofing stapler is specifically designed for applying asphalt or fiberglass roofing shingles. This stapler will drive up to 16-gauge wire, $1^{1}/_{2}$-inch staples. Use 1-inch staples for new work and $1^{1}/_{4}$-inch staples for roofing over the top of one layer of old shingles.

Make sure that the staples are driven flush with the shingle surface, but not through the shingle or less than flush.

Electric-powered nail guns and staple guns are not nearly as expensive to purchase as pneumatic staple guns. Be absolutely certain that the equipment you buy or rent is capable of driving galvanized or aluminum roofing nails or staples that are 1 inch long for one layer of shingles and 1 1/4 inches long for roofing over the top of a layer of old shingles. Never attempt to use brads to install shingles.

The Duo-Fast HT-550 hammer tacker (FIG. 2-17) is ideal for use by professional roofers who must rapidly staple building paper. Such manually operated hammer tackers are especially useful for applying felt to mansards.

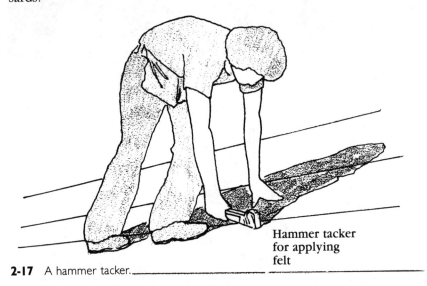

Hammer tacker for applying felt

2-17   A hammer tacker.

Laddeveyor   A *laddeveyor* is a combination ladder and conveyor belt. An aluminum ladder with a gas-driven motor, a pulley, a sled, and a bundle catcher are designed to lift bundles of shingles to one- or two-story roofs (FIG. 2-18). This type of equipment is much too expensive to purchase for small jobs, but it is definitely worth renting if your shingles will not be delivered to the roof surface.

Many shingle suppliers have shingle-delivery equipment. Call the suppliers in your area to find a dealer with such equipment. See FIGS. 3-4 and 3-5. The savings in labor will more than offset any additional charge for delivery of the shingles to the roof.

Ladder   It is essential that the ladder you select be sturdy and the proper length for the job. The four common grades of wood, aluminum, or fiberglass extension ladders are light-duty (household use), medium-duty (commercial grade), heavy-duty (industrial use), and extra-heavy-duty. The grades refer to the safe working loads for each category of ladder. If your shingles will be delivered to the roof surface, a medium-duty ladder—designed to carry 225 pounds—will be more than adequate for use

2-18 A laddeveyor.

by most homeowners. If your weight is above average and you plan to carry the shingles on your shoulder and up the ladder to the roof, plan on using a heavy-duty or extra-heavy-duty ladder. Remember that most shingle bundles weigh at least 75 pounds.

Also, keep in mind that extension ladders are not designed to extend to full length. Because the ladder should extend 3 feet above the eaves, only 21 feet of a 24-foot ladder effectively can be used. Extension ladders are typically sold in 4-foot-length increments from 16 to 40 feet. The owner of a single-story home should select a 16- or 20-foot ladder. In general, a 24- or 28-foot ladder will be a good selection for use on a two-story home.

Along with a wide selection of conventional wood, aluminum, and fiberglass ladders on the market, several types of specialty products recently have been introduced. These versatile scaffolds and ladders are designed to be used in awkward spaces—by employing hinges and folding sections—or to be converted to work platforms. Such specialty devices will prove especially useful if you must shingle a mansard roof. Otherwise, one of the three basic conventional extension ladders will suffice for the do-it-yourselfer.

If your roof has sections where working directly from a ladder (FIGS. 5-37 and 5-38) is the best way to install the shingles, a U-shaped stabilizer device—available for about $20 from some building suppliers—can be quickly attached to a ladder. The stabilizer can be used to free the work space around gutters and to span such obstacles as windows.

While wood ladders are somewhat less likely to mar siding and gutters and are less expensive than aluminum or fiberglass ladders, they are heavier and more difficult to move around. A typical 24-foot wood ladder weighs about 65 pounds and a comparable aluminum ladder weighs 50 pounds. Fiberglass ladders are the most expensive of the three choices.

If you are using a wood ladder that is several years old, be absolutely certain that the ladder rungs will not break when you place your weight—plus the weight of a bundle of shingles on your shoulder—on the ladder. (See FIGS. 2-19 and 2-20.)

Before you carry any shingles to the roof surface, read the section on bundle-stacking methods described in chapter 3. Keep the following

Bearing point	Distance	Length to buy	Working length	Grade	Duty rating
9^1/$_2$"	2^1/$_2$"	16"	13"	Household	200 lbs.
13^1/$_2$"	3^1/$_2$"	20"	17"		
17^1/$_2$"	4^1/$_2$"	24"	21"	Commercial	225 lbs.
21^1/$_2$"	5^1/$_2$"	28"	25"		
25"	6^1/$_2$"	32"	29"	Industrial	250 lbs.
29"	7^1/$_2$"	36"	33"		
32"	8^1/$_2$"	40"	36"		

3"

Length

Bearing point

Distance (¼ length)

2-19 Ladder guidelines.

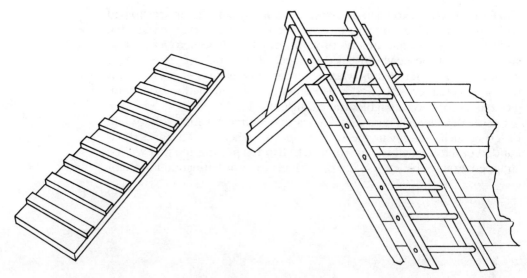

2-20 A "chicken ladder" or a ladder with ridge hooks can be used on steeply pitched roofs.

safety guidelines in mind any time you use a ladder.

- Remember that a ladder can conduct electricity. Stay away from all wires.
- The base of the ladder must rest on an even surface and be placed so that the distance from the eaves is equal to $1/4$ the length of the ladder.
- The ladder should extend about 3 feet above the roof surface.
- Never attempt to use a damaged ladder. Having the rungs of a ladder snap while you are halfway up a two-story climb is not a pleasant experience.
- If you are using an extension ladder, make certain that the locks are fully engaged.
- Wear rubber-soled, slip-resistant shoes.
- When you finish work for the day, take down the ladder and put it out of the reach of children.

Nail Bar   A nail bar (FIG. 2-21) is very handy for removing hard-to-get-at nails and flashing that must be pulled when you are tearing off shingles.

2-21   A nail bar.

Hip Pad   If you are going to do a great deal of shingling, a rubber hip pad will save wear and tear on you and your jeans. As you sit on the roof, the pad fits snuggly around the area of your body that drags across the abrasive surface of the newly installed shingles. A new pair of jeans can be worn out in a matter of days when subjected to this kind of treatment. A hip pad will last for years, and it also provides some insulation from shingles that absorb considerable heat from sunlight.

The McGuire-Nicholas Mfg. Co. is one source for hip pads. When ordering such equipment, be sure to specify whether you are a right-handed or left-handed roofer. For a right-handed person, the pad will be strapped around the left leg; the opposite is true for left-handed roofers.

Chapter 3

Preparations

*T*his chapter explains the preparations that are necessary before you begin nailing three-tab, asphalt fiberglass-based shingles. The general techniques described here can be applied to other styles and shapes of shingles by following these instructions and by noting the information provided by the manufacturer of the roofing material you purchase for your home. Refer to the backs of shingle bundles for specific manufacturer's instructions before you begin to install your shingles.

Examine each section of your roof to determine if the pitch is too steep or if you feel you will be too high off the ground to work with confidence. Now is the time to walk over the roof surface in order to decide if you can do the work of if you should call professional roofing contractors for estimates.

SAFETY GUIDELINES

If you decide to do the work yourself, keep the following safety points in mind.

- Wear comfortable work clothes. Don't wear jewelry or a watch.
- Roofing your home could easily mean lifting several tons of roofing materials. Work at a pace that is compatible with your health and conditioning. Don't overextend yourself.
- During warm weather, take frequent breaks and drink a lot of water.
- Stay off the roof during very hot weather, when it rains, or when it is very windy. Don't attempt to walk on a roof that has frost, ice, or snow on it. Frost is especially dangerous because it isn't always easy to see on a roof.
- Use a very sturdy ladder placed at the proper angle and extended over the eaves by 3 feet. Keep ladders away from power lines. (Refer

to the chapter 2 section on ladders for additional information on ladder selection and use.)

- If you drop something while you are working on the roof, don't attempt to run after it. One more trip down and up the ladder is better than a headlong trip off the roof.

- Don't overload one section of a roof with materials.

- Never walk on felt that is not nailed securely.

- Never leave scraps and shingle wrappers scattered over the roof. Toss trash in one pile on the ground.

- Unless you are already on the ground, don't step back to admire your work.

ESTIMATING MATERIALS

The first step in estimating materials is to determine the total number of square feet of material you will need to cover your roof. The simplest way to find the total area is to climb onto the roof to make careful measurements. Be sure to put on shoes with soft soles, use safe ladder procedures, and take a tape measure, a pencil, and paper with you.

The design of your structure will be a key factor in how simple or difficult it will be to accurately estimate the amount of materials you will need. Simple rectangular shapes—without dormers, valleys, or other obstacles and complications—are computed by first measuring up the *rake* (the parallel edges) of the building, then across the eaves, and then multiplying the two figures. The result is the total square feet of one-half the roof. Twice that number will give you the approximate total square feet for the roof. The following figures can be used as an example for computing the total square feet of a typical, simple rectangular roof surface.

rake:	16	feet
eaves:	×50	feet
subtotal:	800	feet
	×2	
total:	1600	square feet
	(16 squares)	

When valleys, dormers, and hips are part of the structure, things get a bit more complicated. Try to partition large areas of the roof into squares or rectangles to make measuring and estimating easier. Remember that you are making an estimate. You should not try to figure down to the last shingle, the last piece of flashing, and the very last nail. A reasonable guideline is to allow an additional 10 percent for waste material.

In addition to finding the total square feet of the roof surface, you will also have to calculate the amount of material needed for *capping* and for *border shingles*. To determine the number of shingles needed for capping, divide the total one-way length of the eaves measurement plus the length of any hips by 3. Next you must multiply that total by 7. Divide by

3 (because a shingle is 3 feet long) and multiply by 7 (because it takes seven individual caps to cover the length of one installed shingle). Note that each three-tab shingle will produce three cut caps. (For more information on how to prepare capping and install border shingles, see chapter 5.) The following is an example of how to compute capping and border-shingle requirements.

eaves measurement: 50 feet
hip measurement: +0 feet
subtotal: 50 feet

after dividing by three: 17 (round up)
×7
total number of capping shingles: 119 (about 4^1/3 bundles)

Border shingles are applied to the outer edges of the roof to provide additional protection (essential on the bottom course) and to present an even edge, as seen from below, at the rake. To calculate the number of border shingles needed, add each of the rake measurements, plus twice the eaves measurement, and divide that total by 3. Here is an example:

eaves measurement: 50 feet
×2
subtotal: 100 feet

16 feet
×4
subtotal: 72 feet

total: 172 feet

dividing by 3: 57 border shingles

The next step is to add the number of capping shingles to the number of border shingles. The total in this example is 176 shingles. This total is divided by 27 because that is the number of shingles in a typical bundle (this varies with the type and weight of shingles). Use the following as an example:

capping shingles: 119
border shingles: +57
total: 176

divided by 27: 6^1/2 bundles (233 square feet; 3 bundles equal 1 square)

To find the total number of squares for the estimate, add the total square feet of both sides of the building (1600 square feet), the total square feet needed for capping and border shingles (233 square feet), and 10 percent for waste (183 square feet). The total of 2016 square feet means that you should order 20^1/3 squares of shingles from your supplier.

Remember to add the cost of nails, roofing cement, valley flashing, felt, and tools to the estimated total cost. Use the following descriptions of materials as guidelines.

Table 3-1 Recommended Nail Lengths.

Nail lengths	Roofing materials
1″	Asphalt or fiberglass shingles on a new deck
1″	Roll roofing on a new deck
1¼″	Reroofing over old asphalt shingles
1¾″ or 2″	Reroofing over wood shingles

Nails   Use hot-galvanized roofing nails or aluminum roofing nails. They should be 11- or 12-gauge nails with ³/₈-inch heads. To apply three-tab shingles using four nails per shingle, you will need approximately 2 pounds of nails per square of shingles. Nails should penetrate ³/₄ of an inch into the roof deck. (See TABLE 3-1 for recommended nail lengths.)

Roofing Cement   A variety of asphalt coatings, adhesives, and cements are available. Asphalt/plastic-based cement is the most frequently used product for waterproofing around chimneys, vents, and skylights. A gallon or two should be more than adequate for the typical 20-square home. Look for a brand that specifies a plastic base or a rubberized compound. As an alternative to messy-to-use roof cement, the 3M company has developed a sealer that is designed to be pressed in place. This caulk comes in 15-foot, ⁷/₁₆-inch-wide solid tape and is guaranteed for 20 years against cracking and peeling.

Valley and Eaves Flashing   Valleys can be protected with galvanized metal or with a mineral-surface product such as E-Z Roof roll roofing or a rubberized polyethylene, self-adhering membrane. To estimate the amount of metal valley material needed, first go to the roof surface and measure the length of the valley. Be sure to add 6 inches for each overlap if you cannot obtain the metal in one continuous roll. Add another 8 inches for overlap at the ridge and eaves. Use the same procedure for estimating mineral-surface roll roofing. If you do use roll roofing for valley flashing, double the amount. Two layers of roll roofing are needed for adequate protection against leaks in valleys. Metal and mineral-surface roll roofing used for valleys should be at least 36 inches wide.

Step Flashing   If you cannot obtain precut aluminum step flashing, cut 7-×-10-inch pieces and bend each piece in half to make them 7 × 5 inches. Allow for a 2-inch overlap of each piece when you are measuring for step flashing that is to be installed along a wall or chimney.

Pipe Collars   For a reroofing job, carefully remove and reuse all pipe collars. If the collars and shingles were properly installed, the old shingles will have protected the collars from wear. For new work, measure the circumferences of the pipes and purchase metal or polyurethane collars.

Felt   For new construction or where the old shingles have been torn off, use No. 15 saturated felt (building paper) as an underlayment for shingles. A roll of felt covers 400 square feet. Despite what some salesmen claim, there definitely is no need to apply a layer of felt when you are installing a second layer of shingles over the top of a layer of worn shingles.

Drip Edge   For new construction or where shingles have been removed, use medium-weight drip edge along the eaves and rakes. Total the eaves and rake measurements and add 10 percent for overlapping and waste. Drip edge is sold in 10-foot lengths.

Tools   You will need a roofer's hatchet (or a carpenter's hammer for non-purists), a nail pouch or nail apron, a chalk line and a tube of chalk, a utility knife and hook blades (or straight blades if you can't obtain hook blades), a tape measure, a trowel, tin snips, rubber-soled shoes, and a sturdy ladder. If you will be tearing off the old shingles, a flat-edged, tear-off shovel is essential. Do not try to tear off shingles with a garden shovel. If the pitch of the roof is steep, you will also need roof jacks and planking.

Consider using electric-powered nailers or pneumatic-powered staplers and a compressor if such equipment is available for rent. Most inexperienced roofers are lucky to be able to nail one square an hour by hand. A powered staple gun will double your speed.

ROOFING NAIL SELECTION AND USE

To install three-tab, asphalt fiberglass-based shingles on a new deck, use 1-inch galvanized or aluminum roofing nails. (See TABLES 3-2 and 3-3.) To install three-tab shingles over one layer of worn asphalt shingles, use $1^1/4$-inch galvanized or aluminum roofing nails.

If you plan to install shingles over roof-deck insulating building materials such as Homasote or Thermasote, you will need $1^3/4$-inch to $2^1/2$-inch roofing nails. In order for the nails to hold properly, the nails you use should penetrate twice the thickness of the combined insulation, sheathing, and roofing materials.

Longer nails are difficult to drive, and they require more time to install than 1- or $1^1/4$-inch roofing nails. An experienced roofer can drive 1-inch nails with one stroke of a roofing hatchet and $1^1/4$-inch nails with one, or at most, two strokes. Longer nails require several strokes each, which might not seem significant until you consider how many nails it takes to roof a building.

Nails have inch, penny, and gauge designations. The abbreviation *d* stands for the penny sizing of some nails. The term originated in England where handmade nails were sold for pennies per hundred. For example, 100 1-inch nails sold for two pennies; therefore, they became known as 2d nails. The *gauge* of nails relates to the diameter of the wire used to manufacture the nails. Roofing nails are 10 gauge.

Roofing nails are generally sold in 1-pound, 5-pound, and 50-pound boxes. There are approximately 600 aluminum roofing nails per pound, but only about 200—because of their greater weight—galvanized steel

Table 3-2. Galvanized Roofing Nails.

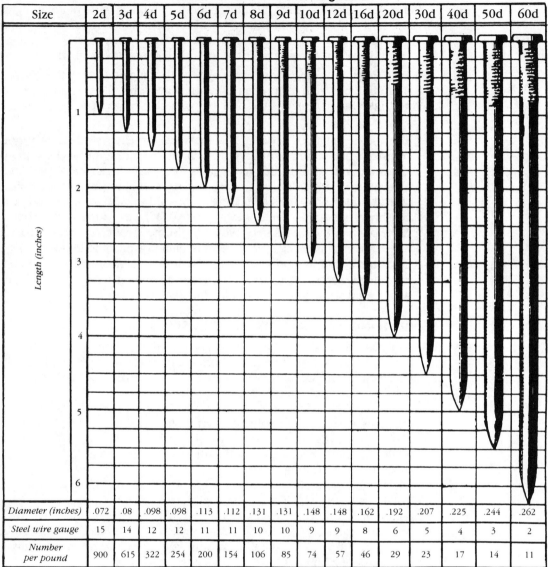

Size	2d	3d	4d	5d	6d	7d	8d	9d	10d	12d	16d	20d	30d	40d	50d	60d
Diameter (inches)	.072	.08	.098	.098	.113	.112	.131	.131	.148	.148	.162	.192	.207	.225	.244	.262
Steel wire gauge	15	14	12	12	11	11	10	10	9	9	8	6	5	4	3	2
Number per pound	900	615	322	254	200	154	106	85	74	57	46	29	23	17	14	11

Length (inches)

Table 3-3. Aluminum Roofing Nails.

Length	Gauge	Head diameter	Nails per pound	Nails per box	Shingle coverage per box
1″	10	7/16″	605	980	300 square feet
1 1/4″	10	7/16″	491	980	300 square feet
1 1/2″	10	7/16″	417	980	300 square feet
1 3/4″	10	7/16″	368	650	200 square feet
2″	10	7/16″	336	650	200 square feet
2 1/2″	10	7/16″	274	325	100 square feet

roofing nails per pound. You will need a little over 200 nails per 100 square feet of roofing area.

▶ **Application Tip** Almost everyone takes for granted that they know the proper way to hold and drive nails. Most people use the technique favored by carpenters. The nail is held near the nail head between your thumb and forefinger. This method is fine for carpenters but for roofers it is too slow. It is also an easy way to pinch your fingers. In addition, it is very difficult to hold more than one nail in your hand when you use this technique.

Nailing shingles with a roofing hatchet is time-consuming and repetitious work. Each full shingle must have four nails. Each time, try to take four to six nails from your nail pouch and drive them before returning your hand for more nails. Hold the nails in the palm of your hand. Cup your hand and use your thumb to "roll" the nails between your fingers. Lay your knuckles against the deck, palm facing up, and practice nailing. The heads of the nails will be easy to tap with the hatchet and you need have little worry about pinching your fingers. It is not important which fingers the nails are rolled between, but it is important that the points of the nails be at the proper angle for setting.

The secret of the technique for rolling nails between your fingers is to position the nail properly and tap the nail head very lightly with your hatchet head so that the nail just remains upright in the shingle. Take your hand away from the nail and drive the nail with one stroke of the hatchet.

The nail head should be flush with the surface of the shingle. Do not drive the nail head deep into the shingle and do not leave it less than flush with the shingle. The idea is to develop a rhythmic motion that allows you to set a nail by tapping it, move your hand, drive the nail, and go to the next nail.

It is crucial that you install nails or staples at the proper locations (FIGS. 3-1 through 3-3). Each shingle has a series of adhesive strips on the face of the shingle. Nails must be driven below these strips, but above the slots that make up the tabs of the shingle. If you drive nails above the adhesive strips, the nails will miss the shingle course underneath and you will not have double-nailed the courses. It is essential to have the double-nailed installation of the shingle courses. On the other hand, nailing too low—below the watermark—will expose the nail heads and leaks will occur.

3-1 Proper nail locations.

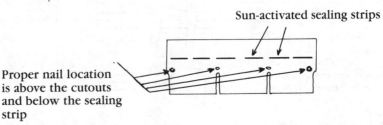

Sun-activated sealing strips

Proper nail location
is above the cutouts
and below the sealing
strip

3-2   When nails are properly located, the shingle courses will be double-nailed.

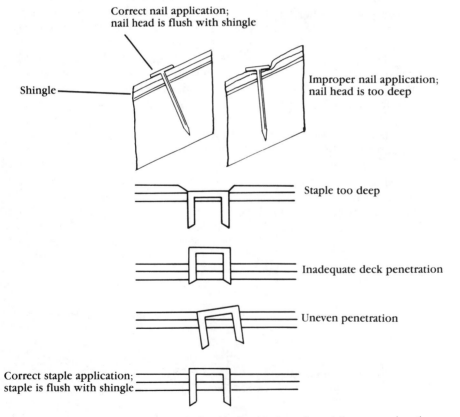

Correct nail application;
nail head is flush with shingle

Shingle

Improper nail application;
nail head is too deep

Staple too deep

Inadequate deck penetration

Uneven penetration

Correct staple application;
staple is flush with shingle

3-3 Nails or staples must be driven flush with the shingle surface at the proper locations.

SHINGLE DELIVERY

Getting the bundles of new shingles on your roof is a task that will require considerable effort. If your health and physical condition are less than excellent, you should not attempt such strenuous work. Although you will be lifting several tons of material, there are several pieces of equipment and several techniques you can use to make the labor easier to accomplish.

Equipment

When you telephone local roofing-material suppliers, you will probably discover that some:

- Have the type of shingle you want, but do not have delivery trucks.
- Will deliver shingles to a job site, but do not have the equipment to place them on the roofs of buildings.
- Will deliver shingles and have the equipment, such as a laddeveyor or conveyor belt, to reach the roofs of one- or two-story structures.

An added cost for shingle delivery to the roof definitely will be worthwhile. In most such cases, the driver of the delivery truck will place the

3-4 If the shingle supplier will deliver the material to the roof, ask the driver of the truck for advice on where to position the boom.

3-5 The shingles are stacked at the eaves because the carpenters were not yet finished sheathing the roof.

bundles on the conveyor belt and you will have to stack the bundles on the roof (FIGS. 3-4 and 3-5).

Bundle-Stacking Methods

When shingles are delivered to the job site, but not to the roof of the building, the ideal storage method is to stack them not more than 4 feet high on wooden pallets. If you will not be carrying the bundles onto the roof immediately, cover the shingles with a tarpaulin. Shingles should not be saturated by rainwater immediately prior to application. Keeping the paper wrappings dry makes carrying the bundles less difficult than if they are wet. The wrappings on wet bundles of shingles will split when you try to lift the bundles. Also, carrying dripping-wet shingles up a ladder makes a difficult task a very unpleasant one.

You should not attempt to carry dozens of bundles of shingles unless you are free of back troubles, have excellent health, and are in excellent condition. When you lift a bundle from a stack on the ground, guide it with one arm so that the shingles will land flat on your shoulder, and let momentum and gravity topple the bundle from the stack to your shoulder.

The bundle should be centered on your shoulder. Use your free arm to grasp the side of the ladder as you climb the rungs until your shoulders are about 2 feet above the eaves. Gently lower the bundle to the roof. By making two more such trips, you will have plenty of material to start work. Carrying too many bundles to the roof at one time is exhausting labor and places too many bundles on only one area of the roof.

As you reduce the number of bundles stacked on the ground, it will become more difficult to get the bundles from the stack to your shoulder. An efficient technique for lifting the bundles is to:

1. Lean the bundle against your thigh.
2. Grasp the sides of the bundle with your hands.
3. Bend your knees slightly.
4. Pull with your arms.
5. Flip up (not end over end) the bundle to your shoulder.

This technique takes some practice before it becomes routine. Don't try it with wet or broken bundles.

If the roof you will be working on is not steep and your roofing-material supplier has the equipment for delivering the shingles to the roof, you should stack the bundles across the peak of the building (FIGS. 3-6 and 3-7). Ask the driver of the delivery truck for advice on positioning the conveyor belt. It will probably be necessary to position several bundles or planks under the top of the conveyor-belt mechanism so that the belt travels freely (FIG. 3-8). It will also make it easier to catch the bundles.

After all the equipment has been set up and the first bundle has been caught, walk to the ridge of the roof and position the bundle so that it is approximately 6 inches below the ridge (FIG. 3-6). The bundle must be placed parallel with the peak. Position the next bundle so that it creates a platform for four or five more bundles. Repeat this pattern until you have stacked all of the shingles on the roof. This technique will allow you to stack the bundles flat; they will not slide off the roof.

If your roof is too steep for stacking bundles, roof jacks and planking will hold several bundles at a time. Do not overload the jacks and planking. Lay the shingles against the roof and use the planking to prevent the bundles from sliding. (See FIGS. 2-15 and 5-49.)

3-6   At the ridge line, position the first two bundles to form a platform.

3-7 Stacking the bundles at the ridge line provides room to work, and the shingles won't slide.

3-8 Raising the boom off the deck allows room for the conveyor mechanism to operate and makes lifting bundles easier.

Chapter **4**

Tear-off Work

Worn shingles should be torn off if they are warped and curled (FIG. 4-1) to such an extent that new shingles cannot be applied evenly over the old shingles. If your home already has two layers of shingles, you must tear off both layers before you install new shingles.

The extra weight of the additional shingle layers is the primary reason why no more than two shingle layers should be applied on a home. The typical residential building is not designed to handle the excessive stress caused by the addition of several more tons of roofing material. Another factor is that standard roofing nails will not adequately penetrate the roof deck through more than two layers of shingles. Some local building codes prohibit the application of more than two layers of shingles because of the difficulty of chopping through several layers of shingles if there is a fire.

Tearing off shingles is a difficult and physically demanding job. Have a sturdy truck available to haul away several tons of material. If possible, park the truck under the eaves and push the material directly from the roof onto the truck bed (FIG. 4-2). Do not attempt to tear off and reshingle more of the roof area than you can manage in one day's work.

Keep in mind the weather forecast for your area. If you give into the temptation to remove as much of the old roof as possible, a rainstorm just might drench everything in the middle of the night. If for some reason you find that you have uncovered more roof area than you can reshingle, carefully nail felt over all of the exposed deck. Make sure that there are no wrinkles or tears in the felt. Nail wood feathering strips parallel with the layers of overlapped felt. Carefully secured felt will provide adequate waterproofing for a short time.

Be sure to wear gloves when you are tearing off shingles. If you aren't careful, blisters can be easily and painfully obtained while doing strenuous tear-off work. Begin tearing off shingles by slipping the edge of the

4-1 Warped and curled shingles must be removed.

4-2 Position the truck directly under the eaves and push the debris into the bed.

shovel under the edge of a few pieces of capping (FIGS. 4-3 and 4-4). After you have removed capping along the length of two or three shingles (FIGS. 4-5 and 4-6), turn and face down the slope. Slip the edge of the shovel under the edge of the top of the first shingle course (FIG. 4-7). Pry the nails loose (FIGS. 4-8 and 4-9), but do not try to pull each individual shingle out one at a time. Instead, work the shovel further down under the second course and pry more nails loose (FIG. 4-10).

Try to loosen as many nails as possible and "roll" several courses of shingle down the slope. Continually pry and push layers of shingles. Work from the sides of the area you have cleared to get at the capping and top course of more shingles on one section of the building. Always begin at the ridge and work down by tearing large swaths of old shingles. Let gravity help you remove the debris. (See FIGS. 4-11 through 4-13.)

Be sure to remove even the smallest pieces of shingles (FIGS. 4-14 and 4-15). A nail bar will prove very useful. Pull all nails or drive them into the deck (FIG. 4-16). Replace any damaged planking (FIG. 4-17).

When you remove the shingles around a vent, work carefully and keep the flange. Flanges almost always can be used again because the old shingles, if they were properly installed, will have protected the flanges from the weather over the years.

4-3 Begin tear-off work at the ridge.

4-4 Slip the edge of the tear-off shovel under several pieces of capping.

4-5 Pry away both layers of shingles and several pieces of capping.

4-6 Remove two or three shingles.

4-7 Loosen the edge of the top layer.

4-8   Pry away at the second layer of shingles.

4-9   Find the nail heads with the flat edge of the shovel and pry.

4-10 Work the shovel under both layers and pry more nails loose.

4-11 Clear the first few courses across the ridge.

4-12   Work down the roof section by tearing large swaths of both layers of worn shingles.

4-13   Allow gravity to help you remove debris.

4-14 Sweep the deck to remove even the smallest pieces of worn shingles and nails.

4-15 A bucket will prove useful for collecting the many small pieces that result when worn shingles are torn from a roof.

4-16   Watch closely for any remaining nails that must be pulled or driven into the deck.

4-17   Wood shingles were torn from the roof of this building. Plywood and felt were installed prior to the laying of fiberglass-based shingles. The height of the building and the pitch of the roof make this a job for professional carpenters and roofers.

LAYING FELT

Use No. 15 saturated felt to provide a smooth surface on which to nail shingles and as temporary waterproofing in case of rain. Lay the courses of felt horizontally and begin at a bottom corner of the roof. Carefully cut off the binding—but don't cut into the roll—and unravel 2 or 3 feet of the roll. Position the roll of felt so that it can be rolled across the bottom of the roof even with the eaves. Kneel and hold the roll in both hands. Maneuver the felt into position so that it covers the deck right up to the edge of the rake and eaves, but not over the sides of the building. (See FIGS. 4-18 and 4-19.)

When you are satisfied that the roll is positioned properly, drive about five nails into the top right-hand corner of the felt. Roll out the felt no more than about halfway across the average-size roof. If there is any wind at all, roll out the felt only about 10 feet. Never walk on felt that has not been nailed down. Pick up the roll with both hands, pull, straighten, and align the felt along the eaves. Be sure that there are no wrinkles. From behind the roll, reach over and nail down the top of the strip with roofing nails spaced evenly 6 to 8 inches apart. Nail the middle and bottom of the felt with a similar series of nails.

Roll out the felt toward the other end of the roof. Leave yourself enough room to pull the felt free of wrinkles and set it even with the eaves. Repeat the nailing pattern and remember not to walk on unnailed felt. Unravel a few more feet and cut it with a utility knife. Trim any felt that overlaps the rake. Nail down the last few feet of the first course of felt.

Start the next course of felt. Be sure to position the roll so that there is a 2-inch overlap of the first course of felt (FIGS. 4-20 and 4-21). Use the white lines printed on the felt as guidelines for lining up the courses. The bottom of the second course of felt should be on top of the first course so that any moisture will flow over the layers of felt.

Roll out a few feet of felt and align the edge of the felt along the rake and the first course (FIG. 4-22). Remember to have a 2-inch overlap. Drive about five nails in the top right-hand corner and roll out the felt to about halfway across the roof. Stand on the first course of felt and work with the second course of felt above you. It is not necessary to completely nail the top edge of the course. A few nails to hold the top in place will be adequate because each top row will become a bottom row once you add another course of felt.

Nail down the bottom and middle of the second course of felt with a pattern of nails every 6 to 8 inches (FIG. 4-23). Remember: never walk on felt that has not been nailed down. Continue laying felt over the remainder of the deck using the same techniques (FIG. 4-24). When you reach the top course, lap about 6 inches of felt over the ridge top (FIG. 4-25). The longer the deck will be exposed to the weather before shingles are applied, the more important it is to lay the felt so that the deck is watertight.

4-18   Position the roll of felt and drive three nails at the top right corner.

4-19   Roll out the felt, align it along the eaves, pull it taut, and drive one nail at the top right corner. Continue across the roof section.

4-20 With a new roll of felt, trim the folded portion before you begin installing the next course.

4-21 With the second layer on top, overlap the courses of felt by 2 inches.

4-22 Roll out the second course of felt and drive a pattern of roof nails.

4-23 Where sections of the felt overlap, drive nails every 3 or 4 inches. Note how the nail is held with the palm of the hand up.

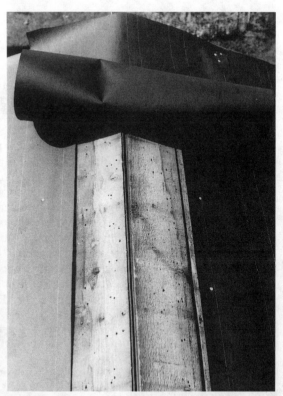

4-24 Continue installing felt courses until you reach the ridge.

4-25 At the ridge, lap the felt over both sides by at least 6 inches.

Using a utility knife, slice any wrinkles and nail the felt so that it is smooth. If the deck will be exposed to the weather overnight or longer, apply a very thin coat of asphalt-based roofing cement to waterproof the areas where cuts have been made.

Do not use roofing cement if you will be reshingling the deck the same day. The cement will take several hours to dry and still might ruin your chalk line when you snap lines over the cemented areas. Also, freshly laid asphalt-based cement would certainly stick to the line and make it impossible to rewind. If there are cemented areas that must have chalk lines over them, nail a piece of scrap felt over the cemented areas. The deck will be waterproof and the chalk line will remain clean.

When you lay felt in a valley, make certain that there are no rips or large wrinkles in the felt. Before you apply the horizontal sections of felt, install a vertical length of felt down the center of the valley. Several sections of felt can be used, but be certain to overlap the higher sections several inches so that water will run over the top of the felt.

Very carefully cut off the horizontal sections of felt at an angle as you reach the center of the valley. If the roof will be exposed overnight or longer, cover nail heads in the valley with a dab of roofing cement.

INSTALLING DRIP EDGE

On homes where the eaves and rakes are visible from the ground, *drip edge* is an attractive way to set off the area where siding and shingles meet. Medium-gauge, aluminum drip edge is inexpensive and very easy to install. Do not use heavy-gauge, galvanized-metal drip edge that is designed for use on hot-tar-and-stone, build-up roofs or drip edge that is so lightweight that it is extremely flimsy. The chief function of drip edge is cosmetic, but it does provide a few inches of additional protection against wind-driven moisture for the area where the deck and eaves meet.

Drip edge is installed with the 3-inch flat portion against the deck and with the creased portion curled snug against the edge of the deck. The 10-foot sections of drip edge are fastened with roofing nails driven into the deck about every 4 feet. Where you overlap drip edge, make sure that the length of drip edge closest to the ridge of the roof is on top to ensure the most attractive appearance when viewed from the ground.

To obtain the best results when you need to cut the drip edge, always first trim the rake edge (FIG. 4-26), set the drip edge in place—without nailing it down—and then use a nail to scratch a line where a cut is needed. Then cut the drip edge, position it, and nail it down.

4-26 Before you install drip edge, use a utility knife to trim the felt at the rake.

Where the rake and eaves meet, use tin snips to cut (from both ends) only the flat portion (FIGS. 4-27 and 4-28) of the drip edge so that it will wrap around a corner (FIGS. 4-29 and 4-30). At the ridge, be sure to make one cut only on the outside (curled) portion of the drip edge (FIGS. 4-31 and 4-32) to eliminate the bulge that would otherwise remain. Drape the remainder of the section over the top of the ridge and down the rake. Be

4-27 To install drip edge at the corner of a rake and eaves, cut only the flat portions of the drip edge.

4-28 To fit a corner, be sure to cut both flat portions of the drip edge.

4-29 At the corner, bend the bottom section of drip edge under the section that will be installed along the rake.

sure to lap the flat section of the drip edge over the top of the next full-length piece of drip edge (FIGS. 4-33 and 4-34).

From the ground, look for any loose-fitting or improperly installed drip edge. Making corrections at this point will be considerably easier than after you have shingled the entire roof.

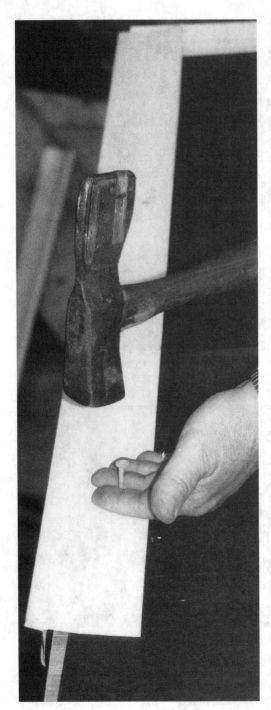

4-31 After you position the drip edge at the ridge and carefully determine where to cut, use tin snips to cut only the outside, curled portion of the drip edge.

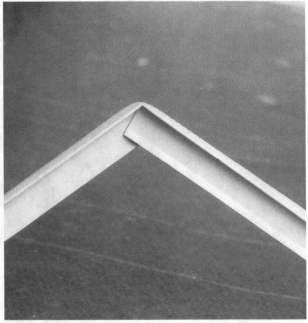

4-30 Install drip edge by driving nails about every 2 feet along the rake.

4-32 Bend the drip edge for a neat appearance at the ridge, and nail the metal in place.

4-33 The top piece of drip edge is lapped over the bottom section.

4-34 Carefully align the drip edge before you nail it in place.

LAYING CHALK LINES FOR THREE-TAB SHINGLES

Vertical and horizontal chalk lines are required for obtaining the proper application of three-tab (36-inch-long) shingles. At the corners and the center of the roof, measure for the bottom course of border shingles by using the following procedures. Extend 1 inch of the tape measure over the edge of the drip edge (or the edge of the eaves if drip edge is not used) and scratch a V in the felt. The point of the V must be exactly at the 12-inch mark. At the corners, make sure that the points of the V marks are scratched within 6 inches of the rakes.

Where you made the mark at the center of the roof, drive a roofing nail about $^1/_2$ inch into the deck at the point of the V. Place the loop that is on the end of the chalk line over the nail head (FIG. 4-35) and walk toward either the right or the left corner of the roof. Unwind the chalk line slowly and do not let it drop or scrape against the deck. Wrap the line around your index finger and pull the line taut. Hold the line against the point of the mark and very lightly snap it once (FIG. 4-36). Rewind the line and walk toward the other corner. Pull out enough line and repeat the snapping procedure.

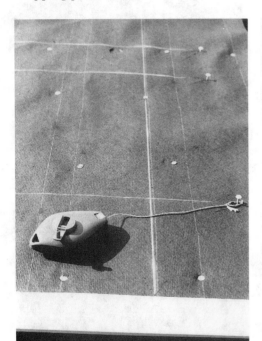

4-35 If you are working alone, you can snap chalk lines by looping one end of the line over the heads of nails driven part of the way into the deck at measured intervals.

4-36 Hold the line against the deck and snap the line once.

Horizontal Lines

Check the width of the shingles you are using. If they are $12^1/_8$ inches wide, measure and mark a V (start from the 12-inch mark where you have just snapped a line) every $5^1/_4$ inches for the first four lines and every $10^1/_4$ inches for additional horizontal lines the remainder of the way up the roof (FIG. 4-37).

If the shingles are 12 inches wide, mark a V every 5 inches, starting from the 12-inch chalk line for the first four lines, and then at every 10

4-37 Snap chalk lines to maintain accurate application patterns.

inches for additional horizontal lines. After the lines are laid, use a tape measure to check the accuracy of at least a few of the lines. Nailing shingles according to inaccurate lines could very easily throw off an otherwise fine-looking job. Done properly, these lines will serve as excellent guides for keeping the shingles straight.

Vertical Lines

As a reference point, the left and right corners in the following description are found by turning your back to the ground and looking toward the peak of the roof. For right-handed persons, start at the lower left corner of the building. For left-handed persons, start at the right corner. In this way, you can keep the work in front of you as you nail the shingles.

Measure in from the rake (with the tape measure extended 1 inch over the drip edge) and scratch Vs at the 12-inch, 30-inch, and 36-inch marks at both the top and bottom (either the right or left) corner of the roof. At the other top and bottom corners, extend the tape measure 1 inch over the edge and scratch Vs at the 12-inch marks. Snap horizontal chalk lines and check them for accuracy. It is extremely important that the 30-inch and 36-inch vertical lines be accurate. Check the measurements for these lines at several places on the roof deck.

If you find that a mistake has been made, you could use a different chalk line with different-colored chalk to snap new lines. When one horizontal line is inaccurate it usually means the others are also inaccurate. As a result, using a different-colored chalk might be the only practical solution. If the vertical lines are inaccurate, and you don't want to use different-colored chalk, take a nail and make a series of scratches in the felt every few inches along the length of the line. Then snap new lines. The scratches will help you avoid getting the lines confused even though the lines are all the same color.

Chapter **5**

Application Patterns

*T*here are three commonly used patterns for applying three-tab, asphalt fiberglass-based shingles. Shingle manufacturers usually specify that either the *45 pattern* or the *random-spacing pattern* be used in order to ensure the best distribution of shingle color blend.

The *straight pattern* is often used to shingle dormers, catwalks, and other roof areas that are primarily vertically oriented surfaces. In order to shingle A-frame buildings and other roofs that are very steep, you must use roof jacks. It is much easier to use the straight pattern when you are working with roof jacks because the work is in front of you and fewer movements of the planks and jacks are required.

STARTING POINTS

If the roof sections of your home consist only of squares and rectangles, selecting a starting point is easy. Working on the back portion of the roof first will give you the opportunity to become familiar with the application pattern and the time and effort needed to do the job. If the shingle cutouts are not perfectly aligned for your first attempt, it will not be especially visible on the back of the building.

If you are right-handed, begin at the lower left corner of the back of the roof in order to keep the work in front of you. If you are left-handed, start work at the lower right corner of the back of the roof. The descriptions accompanying FIGS. 5-1 through FIG. 5-25 are oriented for right-handed persons in order to simplify the instructions. Reverse the starting points—to the opposite side of the roof sections—if you are left-handed.

The starting-point suggestions shown in FIGS. 5-1 through 5-25 will help you deal with such intrusions in the roof lines as dormers, valleys, and walls. Depending on how you proceed, the sequence in which you shingle the sections of your roof will make the work easier or more difficult.

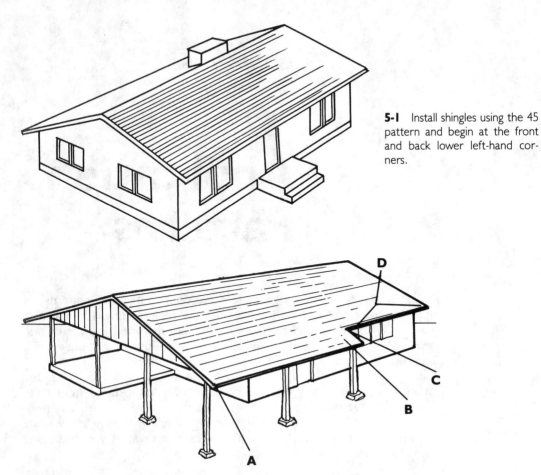

5-1 Install shingles using the 45 pattern and begin at the front and back lower left-hand corners.

5-2 Begin at the left-hand corner (A) and shingle to the angle (B). Be sure to install border shingles at the angle and continue the 45 pattern past the break in the roof line (C). If a short course is needed to maintain the proper exposure, make the adjustment at the course (D) that aligns with the break.

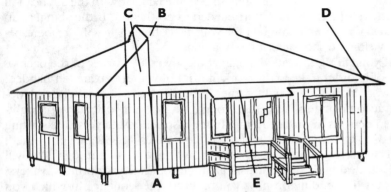

5-3 Snap chalk lines between points A and B. Fill in section C. Shingle across to point D but high-nail and back down across point E.

5-4 Beginning at point A, use the 45 pattern to shingle past the angle, as shown in Fig. 5-2. On the second story, begin at point B. When you measure for chalk lines, carefully take into account the chimney. If a chimney is at the ridge line (C)—where a left-hander would begin this section—measure for the vertical chalk lines just below the chimney and draw the lines through the marks so that they extend past the chimney to the ridge.

5-5 Begin at A, snap tie-in chalk lines at the dormers (B), and shingle the dormer tops by starting at the rakes (C).

5-6 Starting at point A, shingle to the wall (B) and install step flashing. Start the second section at point C.

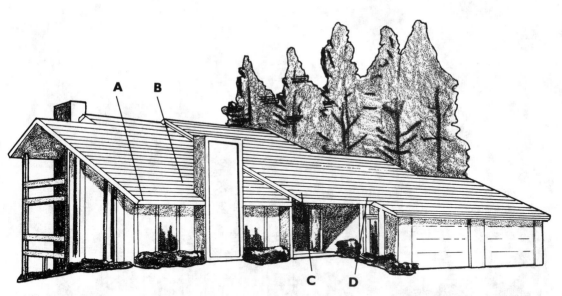

5-7 At point A, begin shingling and use chalk lines to tie in the pattern around the chimney (B). Shingle across from point C and back down the pattern at point D by high-nailing the course (C).

5-8 Start each of the three sections at the lower, left-hand corners. Make certain section C is not thrown off by a wrong measurement at the wall.

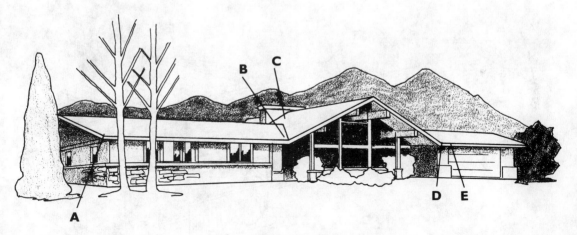

5-9 Begin at point A and shingle toward the valley at point B. Work up the rake and toward the valley and chimney. At point C, back the shingle pattern down the roof. Start at the rake to shingle point D. At point E, start at the eaves, square off the section with chalk lines, and back in the pattern toward the valley.

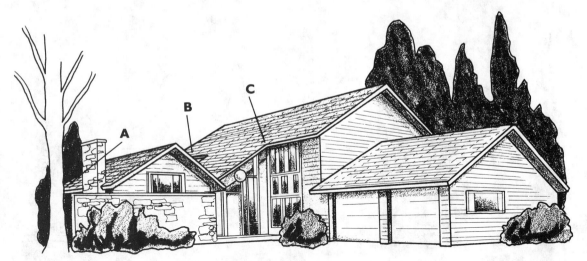

5-10   Begin at point A and tie in past the chimney. Shingle up the rake (B) and high-nail across to point C. Back down the bottom section of the main roof.

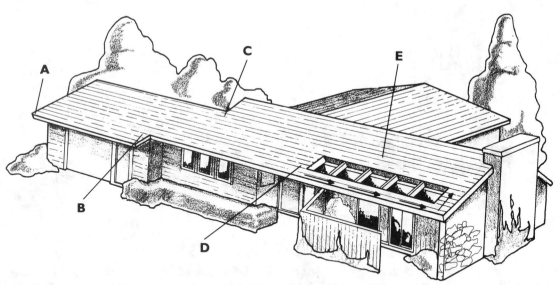

5-11   Using the 45 pattern, begin at point A and shingle past the first angle. At point B, install border shingles along the rake and eaves, and lay a short course if necessary. At point C, snap chalk lines in order to continue the pattern. At point D, shingle past the bottom of the obstacle and up the rake. Snap chalk lines across point E to continue the shingle pattern.

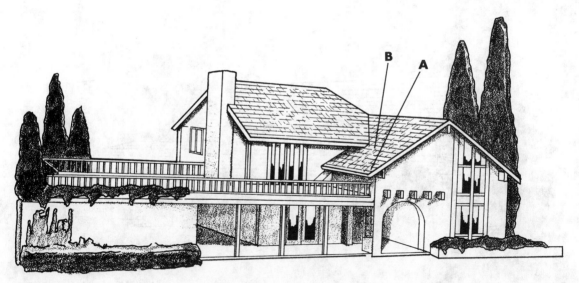

5-12 Because of the valley and the wall, begin shingling at point A. Where the valley intersects (B), the valley material must go on top of at least one course of shingles.

5-13 Use roof jacks and planking on a roof that has a pitch this steep. Begin at the lower left corners of each section (A, C, D). Use roofing cement at the tabs to secure the shingles where the roof line steepens (B).

5-14   Shingle the top section first (A). Square off the hip section (B) and shingle across the roof using the 45 pattern.

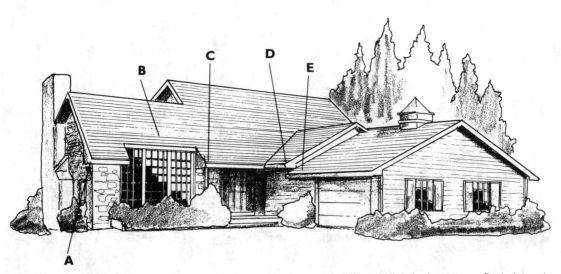

5-15   Use the straight pattern to begin at point A. At point B, snap chalk lines and tie in the pattern. Back down the courses to the eaves and shingle into the valley. Begin shingling at the rake at point D and at the eaves and wall at E.

5-16 Snap vertical chalk lines between points A and B. At point C, install a horizontal chalk line and border shingles across the top of the break in the roof line. Back down the shingle pattern at point D.

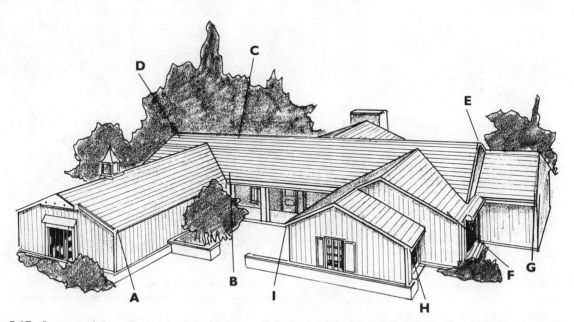

5-17 Because of the valleys and rakes, this home design looks difficult to roof. Dividing the job into sections will make the work manageable. Start at point A and shingle toward the valley. Snap chalk lines between points B and C and fill in across to points D and E. Back down along the rakes and continue to fill in. Start at the rakes for the remaining sections (F through I).

5-18 At first glance, it looks like you should begin shingling this roof design at the middle of the house. However, starting at the middle would require considerable backing in. By starting at the left rake (A), you can keep the work in front of you. Be sure to install border shingles along the eaves and rakes where the roof lines are angled (B, C, and D).

5-19 Shingle the entire top portion (A) of the building first so that you will not mar the shingles with a ladder extending from the lower section to the top section. Shingle the lower section starting at point B so that the pattern extends across and up the roof line (C).

5-20   Although it means backing in, always work from the eaves toward the valley of this type of house design. Because the roof sections are not large, consider using the straight pattern (A, B). At point C, the bottom of the right side of the valley material must extend over the top of at least one course of shingles.

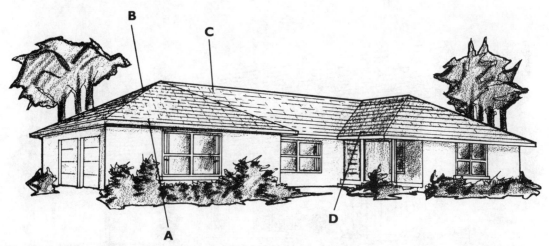

5-21   Hip roofs require the cutting of a great many shingles as you install the pattern. Where possible, square off the sections with chalk lines (A, B, C, and D).

5-22 Begin work at the rakes (A, B). At the valley (C), install the bottom left corner of the valley material over the top of at least one course of shingles.

5-23 Start shingling at point A and continue the border shingles and the pattern past the angled section (B).

5-24 This home design would be easy to shingle if the pitch of the roof were not so steep. Roof jacks will be needed to do the work. Working from a ladder, begin at point A and shingle across the eaves. Install as many courses as possible from the ladder.

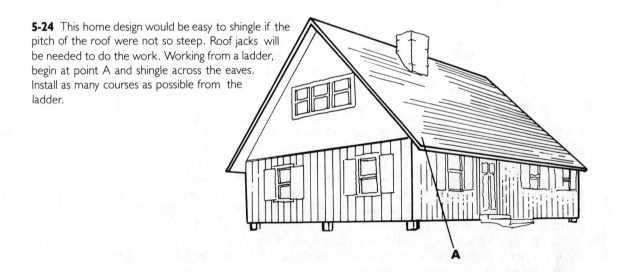

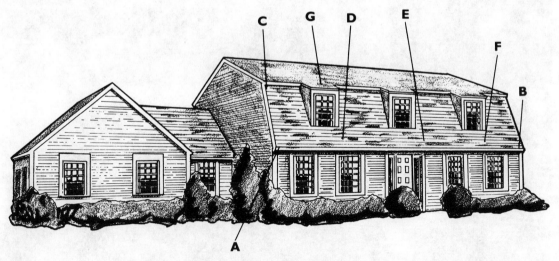

5-25 Because mansard-roof shingles are highly visible, consider having a professional roofer do the mansard work while you do the remainder of the roof. The mansard should be shingled, starting at point A, across the eaves (B). Continue the pattern up the rake (C) and the dormer walls. Snap vertical chalk lines (D, E, and F) in order to properly continue the pattern. On the top section of the roof, shingle across the eaves and back down the pattern at the dormer tops (G).

SHORTCUTS AND BACK-SAVING TIPS

Here are a few techniques that will save you extra effort and movement
when you are installing shingles. The first few shingles that you install will
have to be nailed into position while you are crouched in a somewhat
awkward position facing the ground (FIG. 5-26). An alternative method is to
work from a ladder (FIGS. 5-27 and 5-28). Be extremely cautious if you prefer
to work from a ladder.

After the first few courses are on, you should sit on the deck with the
work in front of you. If you are right-handed, tuck your left foot underneath your body and balance yourself so that you are comfortable. The
center of your weight should be on your left hip. This position will give
your upper torso freedom of movement. You should be able to nail at
least four shingles before you have to move up the roof. After a little practice, you will get the hang of positioning yourself so that you will be close
to the work but not too close.

Each full shingle must be nailed with four nails. When you are using
the 45 pattern or the random-spacing pattern, there is a technique you
can use to reduce movement and unnecessary stretching. As you sit nailing shingles, you will soon discover that the outside nail on the pattern is
the most difficult to reach. Instead of stretching for it, leave that outside

5-26 Face the ground while nailing the first few courses at the rake.

5-27 The first few courses can be installed from a ladder.

5-28 Don't attempt to reach farther than your arm's length when working from a ladder.

tab temporarily unnailed and catch it on the next run. Each time you will be driving three nails into the shingle that you position and the fourth nail is driven into the tab at your left. Any nails that remain in your hand can be driven into the next few "fourth" tabs of the above courses.

Inexperienced roofers will sometimes contort into all sorts of uncomfortable stances while they try to nail shingles. Even some crews of professional shinglers—fighting gravity and common sense all the way—will work the entire roof from bottom to top with their backs to the peak and each course of shingles below their feet. Such positions will place considerable strain on your back. Don't try to kneel, bend over while standing flat-footed, or work the entire job "upside down."

OVER-THE-TOP SHINGLING

A roof that has one layer of worn but not badly warped or curled shingles can be reroofed by nailing a new layer of shingles over the old shingles. Use a utility knife to cut back the shingles only along the rakes to expose about 3 inches of the deck so just one layer of shingles will be seen from the ground after the new roof has been installed. As a result, the final appearance of the roof will be more attractive than if two layers could be seen at the rake.

It is also very important to remove the tabs on only the third course of the old shingles. Remove only the tabs (FIG. 5-29); do not remove the

5-29 Where you shingle over the top of one worn layer, tear off the tabs of the third course of the old shingles. Use a trowel to help in prying off tabs.

entire course of shingles when you are shingling over-the-top. Done properly, removing the tabs will prevent the buildup of a hump along the third course when the new shingles are installed. If you do not remove the tabs at the third course, the hump will appear because of the layers of border shingles and the additional courses.

Make measurements and lay vertical chalk lines as described in chapter 4. Instead of laying horizontal chalk lines, it might be practical for you to butt the tops of the new shingles against the bottoms of the old shingles as you shingle the roof. In this way, you can follow the pattern set by the old shingles, and it is an easy way to keep the new shingles straight. If you use this method, be certain that the proper 5-inch exposure is maintained and that the new shingles are exactly the same size as the old shingles.

The 45 Pattern

The following instructions are for shingling square and rectangular roof sections. For guidelines on shingling angled roof sections, see this chapter's section on starting points and chapter 6.

To begin nailing shingles using the 45 pattern (FIG. 5-30), first refer to the section on laying vertical chalk lines in chapter 4. After you have

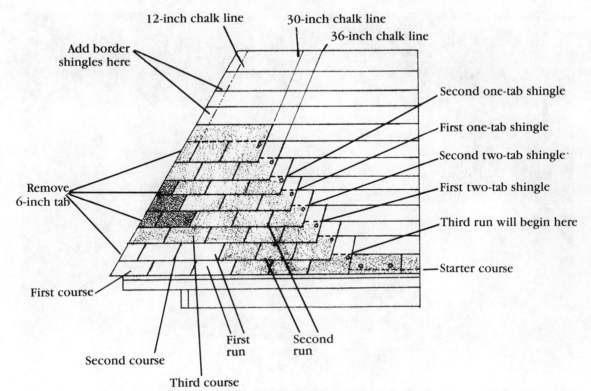

5-30 The shaded shingles highlight the 45 pattern. Maintain the pattern by inserting one-tab shingles, two-tab shingles, and shingles with 6-inch tabs cut from the rake side.

snapped all of the lines, take one bundle of shingles to the corner of the roof where you have snapped the 12-inch vertical lines. Nail the first eaves and rake border shingles (FIGS. 5-31 and 5-32). Be sure to position the shingle so that the top, solid-colored portion of the shingle (granules up) faces outward (toward the ground) and the tabs face toward the roof. The border shingles must be aligned exactly with the chalk lines and be positioned so that they have a 1-inch overhang at the eaves and a 1-inch overhang at the rake.

▶ **Application Tip**   It is best to get into a habit of nailing border shingles above the adhesive sealing strip. Otherwise, you might easily drive a nail where a cutout of the first course of shingles would expose the nail head. This situation would lead to the development of a leak.

If you have installed the border shingles properly, there should be about 6 inches of upside-down shingle visible at the eaves so that the border shingle and the first-course shingle overlap rather than match up. A match-up is a common error that is easy to make and the result would be a series of leaks across the eaves.

First Course   With one vertical border shingle and one horizontal border shingle in place, begin the first course of shingles by positioning a full shingle so that it is aligned with the 36-inch chalk line, and so that it

5-31   The first eaves border shingle is installed so that it overhangs the drip edge by 1 inch and so that it offsets the vertical 36-inch chalk line by 6 inches.

5-32 Install the first rake border shingle so that it overhangs the drip edge by 1 inch at the rake and at the eaves. In order to show the proper alignment, the eaves border shingle (Fig. 5-31) is not shown installed.

extends over the rake by 1 inch. This first course goes directly on top of the border shingles (FIG. 5-33).

Second Course Align the first shingle of the second course. Position a full shingle so that it extends 6 inches over the rake (FIG. 5-34). Carefully align the factory edge of the shingle with the 30-inch vertical chalk line. As you continue installing the pattern up the roof, the shingles will overlap so that a 6-inch tab extends over the rake.

▶ **Application Tip** Most professional roofers would probably let all of the overhanging tabs remain, nail all of the shingles on one section of the roof in place, and then cut off all the tabs from the overhanging shingles at once. Figures 5-35 and 5-36 show how professional roofers use a chalk line to help them trim a rake evenly. Trimming the rake in this manner can be a particularly difficult task for someone unaccustomed to handling a utility knife. The results—especially with sun-baked shingles—can be considerable aggravation, scraped knuckles, and a rake that looks terrible. For do-it-yourselfers, there is an easier way to cut a rake.

Almost all weights and brands of three-tab asphalt fiberglass-based shingles have factory cuts at the 6-inch marks at the tops of the shingles. This is a very convenient and accurate starting point for cutting off the tabs. While the easiest tool with which to make cuts is a pair of tin snips, professional roofers use a utility knife with a hook blade or a straight

5-33   The first course is installed aligned with the 36-inch chalk line and even with the border shingles.

5-34   Align a full shingle with the 30-inch chalk line. The bottom of the second-course shingle must be aligned with the tops of the first-course watermarks. The 6-inch overhanging tab can be trimmed before or after the shingle is installed.

5-35 (Left) A chalk line can be used to mark the proper 1-inch overhang.

5-36 (Above) Trimming the rake with a utility knife is a difficult task for a beginner. Tin snips can be used for this job.

blade because this method is faster. Many roofers prefer to use hook blades to cut asphalt fiberglass-based shingles because straight blades tend to "slide" over the shingle. The hook blade will "grab" the shingle and give a better cutting action.

Whether you use tin snips or a utility knife, first position (but do not nail) the shingle and then mark the shingle where it is to be cut. Then turn over the shingle so that the granules face down. Be careful when you do this type of work because it is very easy to cut all the way through a shingle and into whatever is underneath. Find the factory cut on the proper end of the shingle and cut off the tab. Be careful not to cut off the wrong end of the shingle. If you happen to make a wrong cut (something that is not unusual for beginners), put the scraps aside. You will undoubtedly be able to use them at the other end of the roof. Remember to first position the shingle, and then trim the overhanging tabs as you install the course of shingles along the rake. If you follow this method, mistakes will be less troublesome to correct.

Third Course Position a shingle (minus a ⅓ tab) so that it aligns with the rake (with a 1-inch overhang), and so that it is even with the two courses of shingles below. The cutouts must align precisely. Nail the first ⅔ shingle in place; save the second ⅔ shingle for the fourth course. In doing this step, it is easiest to first take two full shingles and cut off ⅓ tabs (so-called one-tabs) from the rake side (FIG. 5-37) of both shingles. Save the ⅓ tabs for use during the next run. The reason for saving the tabs is explained on page 81.

Fourth Course Begin the fourth course of shingles by positioning the second of the ⅔ shingles with the cut end along the rake. The factory edge must be aligned with the 6-inch factory cut of the shingle you have nailed in place for the third course. (See FIG. 5-38.)

Fifth Course Cut the 6-inch tab from the rake side of the second ⅔ shingle. Carefully align and position this shingle for the fifth course. You should be able to clearly see the shingles taking on a 45-degree pattern of

5-37 Cut a one-tab for later use.

5-38 Align the first shingle of the third course.

"steps" up the roof section. The cutouts must all be aligned in the proper 6-inch pattern. It is essential that these first few courses are properly positioned. An error can easily be fixed at this stage. If you are not absolutely sure that the pattern is correct, stop and make the necessary adjustments.

Remember to install border shingles (FIG. 5-39) and then begin the next run of the pattern (FIG. 5-40). By nailing a border shingle and four full shingles in place, you will reach the top of the pattern. In order to extend the stepped pattern further up the roof, nail a full shingle even with the rake. Cut a 6-inch scrap piece from the rake side of another full shingle (save the scrap for later), and nail the shingle in place.

At this point, retrieve the one-tabs you saved from the first run. Nail the two one-tabs in place to retain the step pattern of the shingles. Continue the pattern by cutting two-tab shingles and 6-inch scraps from the appropriate shingles, as you did for the first run. Shingling the remainder of the section of the roof is a matter of repeating this pattern on the remaining runs. (See FIGS. 5-41 through 5-48.)

5-39 Another border shingle is installed along the rake.

5-40 The tabs of the third course of the worn shingles have been removed. Install a full shingle.

5-41 (Above left) The second shingle of the third course is installed.

5-42 (Above right) Tabs are easier to cut when the slots face away from you.

5-43 (Left) Installing the first shingle of the fifth course.

5-44 Two-tab shingles must be cut and installed on the second run in order to continue the 45 pattern.

5-45 Two runs of shingles have been installed.

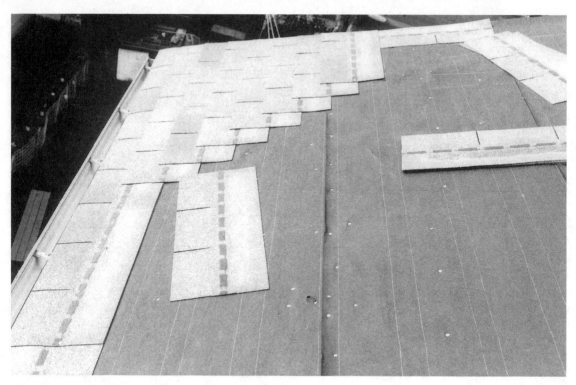

5-46 The border shingle and the first shingle of the third course have been installed. Continue the pattern by installing full shingles.

5-47 With the 45 pattern, the factory cutouts must be aligned on every other course.

5-48 A few more runs will take the pattern to the ridge.

The Random-Spacing Pattern

The techniques for laying asphalt fiberglass-based shingles in the random-spacing pattern are almost identical to the methods used for shingling with the 45 pattern. With the random-spacing pattern, the cutouts on the shingles will align every six shingles. With the random-spacing pattern, shingle cutouts will help reduce wear patterns from rain as the shingles age. Because of this, the shingles will last slightly longer than identical shingles installed in the 45 pattern or the straight pattern.

Begin the random-spacing pattern by snapping chalk lines as described in chapter 4. After you have snapped all of the lines, take a bundle of shingles to the corner of the roof where you have snapped the 12-inch vertical lines. Nail the first vertical border shingle. As with the 45 pattern, be sure to position the shingle so that the top, solid-colored portion of the shingle faces outward and the tabs face toward the roof.

The border shingle must be aligned exactly with the chalk line and be positioned so that it has a 1-inch overhang at the bottom and a 1-inch overhang at the rake. Now position a horizontal border shingle. Turn this shingle so that the tabs face you. Align the shingle exactly with the horizontal chalk line and nail the shingle in place.

Nail the border shingles above the adhesive sealing strips on the shingles. Do not drive a nail where a cutout of the first course of shingles would expose nail heads, causing a leak.

First Course   Now comes the difference between the 45 pattern and the random-spacing pattern. Instead of beginning the first course of shingles by positioning a full shingle so that it is aligned with the 36-inch chalk line and so that it extends 1 inch over the rake, position the first-course shingle so that it extends 3 inches over the rake. Nail the shingle in place and trim the overhanging piece even with the border shingle (leaving a 1-inch overhang).

Second Course   Align the first shingle of the second course by positioning it so that 9 inches of the shingle overhangs the rake and the other side of the shingle aligns with the 6-inch factory cut on the first-course shingle. Trim the overhanging piece.

The 30-inch and the 36-inch chalk lines will not intersect the shingles until several courses in the random-spacing pattern have been installed. Do not match up the bottom border shingles and the first course.

Third Course   Begin the third course of shingles by positioning a two-tab shingle with a portion of the shingle overhanging the rake and the other end carefully aligned with the 6-inch factory cut of the second-course shingle. In other words, the third course must be placed so that it continues the 45-degree step pattern.

Fourth Course   Position another two-tab shingle for the fourth course by aligning it with the 6-inch factory cut of the third-course shingle; continue the step pattern. Trim the overhanging piece.

Fifth Course   Install border shingles and begin a run of full shingles from the bottom. When you reach the top of the pattern, install one-tabs to extend the stepped pattern.

The Straight Pattern

Nailing shingles using the straight pattern aligns the cutouts 6 inches apart on every other shingle. It is best to use this pattern on steep surfaces requiring the use of roof jacks (FIG. 5-49), on roof dormers, and on short, vertical surfaces such as catwalks. The pattern consists of repeating the vertical runs up the roof until the section of the roof has been completely covered. Begin by laying chalk lines as described in chapter 4.

After snapping all of the chalk lines, take one bundle of shingles to the corner of the roof where you have snapped the 12-inch vertical lines. Nail in place the first horizontal border shingle (FIG. 5-31). Be certain to position the shingle so that the top, solid-colored portion of the shingle (granules up) faces outward (toward the ground) and the tabs face toward the roof. The border shingles must be aligned exactly with the chalk line and be positioned so that it has a 1-inch overhang at the bottom and a 1-inch overhang at the rake.

The next step is to position a vertical border shingle. Turn this shingle so that the tabs face you, align the shingle exactly with the horizontal chalk line, and nail the shingle in place (FIG. 5-32).

➤   **Application Tip**   Make a habit of nailing the border shingles above the adhesive sealing strip. Otherwise, you might easily drive a nail where

5-49   The straight pattern should be used on steep roofs that require roof jacks.

a cutout of the first course of shingles would expose the nail head. An exposed nail head would cause a leak.

First Course   With one vertical and one horizontal border shingle in place (FIG. 5-50), begin the first course of shingles by positioning a full shingle so that it is aligned with the 30-inch chalk line, and so that it extends over the rake. This first course goes directly on top of the border shingles (FIG. 5-51). If you have installed the border shingles properly, there should be about 6 inches of upside-down shingle visible on the outside of the vertical chalk lines (FIG. 5-50); the border shingle and the first-course shingle overlap rather than match up. A match-up is a common error that is easy to make, and the result would be a series of leaks across the eaves.

Second Course   The next step is to align the first shingle of the second course. Position a full shingle so that it is even with the rake. Carefully align the factory edge of the shingle with the 36-inch vertical chalk line.

Now you have a choice of cutting off the 6-inch scrap or temporarily leaving the overhanging tab in place until the run has been completed. Most professional roofers would probably let the overhanging tabs remain in place and then cut off all of them at once. Cutting the overhanging tabs can be a particularly difficult task for someone unaccustomed to handling a utility knife. It is easier to use a pair of tin snips. If you do prefer to use a utility knife, a hook blade is easier to use than a straight blade. Trim the

5-50   For the straight pattern, rake and eaves border shingles have been installed.

second-course shingle so that it is even with the vertical border shingle.

Third Course   Begin the third course of shingles by positioning a full shingle so that it is aligned with the 30-inch chalk line and so that it extends over the rake. This shingle must be aligned so that the bottom of it is even with the cutouts of the second-course shingle. (See FIG. 5-52.)

Continue the straight pattern by installing shingles in the vertical run until you reach the top of the section of roof (FIG. 5-53). Every other shingle will overlap the rake by 6 inches and the alternating full shingles will align with the border shingles at the rake. A shingle can be used as a straightedge (FIG. 5-54) to help you trim the 6-inch tab from the second shingle in the starter course. Use the factory cut as a guide.

Be absolutely certain to nail each shingle with four nails. In order to do this when installing the straight pattern, you must lift the tabs on every other shingle course as you install each of the runs following the first run.

➡ **Application Tip**   While you are installing shingles in the straight pattern, you can save time and effort by using the following techniques for positioning bundles of shingles. If the bundles of shingles do not slide as a result of the steepness of your roof deck, place a bundle on the run of shingles you have just installed. The way you position the bundle is crucial to obtaining the greatest shingle-application efficiency. Position the shingles with the tabs facing ''up'' and the cutouts facing toward the deck. The bundle must be close enough to reach, but it must not prevent you from lifting the tabs of the run you have just installed.

5-51 The first shingle in the first course of the straight pattern must be installed so that a 6-inch tab overhangs the rake.

In order to position a partial bundle of shingles that otherwise might slide away on a slightly steeper deck, you can temporarily insert a piece of step flashing under one of the shingle tabs of the straight-pattern run just installed. Lift a tab and gently slide one edge of two or three L-shaped pieces of step flashing snug against the nail holding the shingle in place. The idea is to create a temporary resting place for the loose shingles that are about to be installed. Again, the shingles must be within reach but they must not prevent you from lifting the shingle tabs as you install the current run of the straight pattern. When you are finished with the run, be careful not to damage the shingle tab as you remove the step flashing.

5-52 Align the shingles with the 30- and 36-inch vertical chalk lines. Also, the bottoms of the shingles must align with the watermarks of the course below.

5-53 Four courses of the straight pattern have been installed.

5-54 In order to cut 6-inch tabs from the alternating shingles in the starter course, you can use a full shingle as a straightedge.

CAPPING

Hips and ridges are shingled with capping. If your home design includes hip sections, always finish capping the hips before you cap the ridges. Determine the number of caps you will need by referring to the section on estimating materials in chapter 3.

The fastest way to cut capping from bundles of three-tab, asphalt fiberglass-based shingles is to first turn the bundles granular-side down and tear away the wrapping paper. Cut only one bundle of capping at a time in case you have overestimated the amount of capping you will need. Work on your knees and position the cutouts facing away from you. Begin at either side of the bundle. Steady the top shingle with the palm of the hand that is not holding the utility knife.

Using a hook-blade utility knife, cut off a diagonal slice of shingle beginning at the watermark and going all the way to the top of the shingle (FIGS. 5-55 and 5-56). Work slowly with the a steady pulling pressure on the knife until you get the hang of it. Soon you should be able to make steady, successive slicing motions. Always be very careful not to scrape your knuckles. Don't jerk the knife.

Make the next cut starting at the second watermark. Angle the cut so that you are slicing toward your body. The object is to form a diamond-shaped cap.

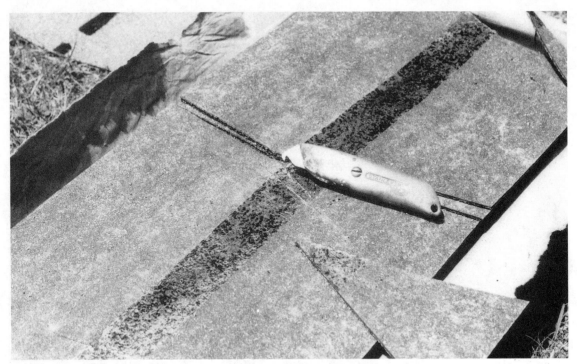

5-55 To cut capping from full shingles, first turn the bundle face down. A hook-blade utility knife or tin snips can be used to cut diagonal scraps from the watermarks to the top of the shingle.

Cut the remaining two sides from the shingle by slicing the scrap from the shingle. Cut the diamond shape for all of the caps. When installed, diamond-shaped caps are much more attractive than square-cut capping.

➤ **Application Tip** Here is a simple and accurate way to lay capping. Use a chalk line and three caps to determine if you have the proper coverage. Position one cap at each end of the roof and place one cap at the center of the hip or ridge. Position the center and end caps so that they straddle the hip or ridge and cover the watermarks on both the side courses or the top courses.

If you are working alone, tap in a nail at the bottom corner of the street side of the center cap. Hook your chalk line on the nail head and walk to either end of the roof. If you are working with others, have someone hold one end of the line in position.

Align the chalk line with the bottom of the end cap. Snap a line, rewind the chalk line, snap another line at the other end of the roof (aligned with the other cap), and check to see if the line is below all of the 5-inch watermarks on the side courses or the top courses of shingles. If the chalk line shows that you will have adequate coverage, you can begin nailing caps. Otherwise, you will have to install another course of shingles. For easier installation, bend but do not crease the caps prior to positioning them.

5-56 Trim both sides of the one tabs to make diamond-shaped capping.

Hip Capping

Because the last hip caps should go under the ridge caps, cap all hips be-fore you install ridge capping. Before you install hip capping, make cer-tain that all of the shingle courses on both roof sections adequately cover the hip. Snap a chalk line using the techniques described in the preceding section.

When you are ready to install the first hip cap at the eaves, position the cap so that the bottom center of the granular portion is even with the 1-inch overhang. You will find that the edges of the cap overlap the bot-tom shingle course on both sections of the roof. Trim this portion of the cap at an angle so that the cap is even with the bottom courses.

Install the caps on the hip by aligning the granular side even with the first cap's 5-inch watermark and the chalk line. Drive a nail into the cap about 1 inch up from the chalk line and between the adhesive strip on the cap and the 5-inch watermark. Drive another nail at the opposite location on the other side of the hip.

Install the second cap so that it aligns with the chalk line and overlaps the first cap. Each additional cap should provide 5 inches of coverage. The caps must overlap so that each cap is double-nailed.

After you have nailed several caps, you will have room to turn your back to the roof edge and sit facing the ridge. If you are working alone, hold about 20 caps in your lap and work up the hip. If help is available, have someone hold each cap in place, even with the chalk line, while you nail. Your helper should sit facing you.

When you reach the end of the hip, cut the last cap so that it "folds" over the ridge. Cover any exposed nail heads with roofing cement.

Ridge Capping

Before you install the ridge caps, make sure that the top courses of shingles provide adequate coverage on both sides of the ridge. The caps must cover the 5-inch watermarks on both top courses across the entire ridge. (See FIGS. 5-57 and 5-58).

To install ridge caps, the traditional procedure is to place the unnailed ends of the caps opposite the prevailing winds. Begin at one end of the roof rather than in the middle.

Align the first cap on the ridge with the granular side even with the 1-inch overhang (FIG. 5-58). Also make certain that the cap is even with the chalk line. Drive a nail into the cap about 1 inch up from the chalk line and between the adhesive strip on the cap and the 5-inch watermark.

5-57   Trim the top of the last course of shingles. The tops of the opposite, last course can be overlapped and nailed or trimmed.

5-58 Align the first cap along the ridge.

Drive another nail at the opposite location on the other side of the ridge.

Install the second cap so that it aligns with the chalk line and overlaps the first cap. Each additional cap should provide 5 inches of coverage. The caps must overlap so that each cap is double-nailed. (See FIGS. 5-59 through 5-63.)

When you reach the end of the ridge, trim the last cap even with the 1-inch coverage so that it shows only the granular portion of the cap. The last cap most likely will be less than a full cap. Drive four nails into the last cap and cover the nail heads with roofing cement.

If the roof has hips, the first and last ridge caps must be cut so that they overlap the hip sections. Cut the end caps so that they "fold" over the hip sections. Cover any exposed nail heads with roofing cement.

Dormer Ridge Capping

To install caps on a dormer ridge, follow the basic instructions outlined in the preceding sections. When you reach the valley section of the ridge, make certain that the last few caps go under the shingles at the top of the valley.

Be extremely careful when you nail the caps as you approach the valley. Use common sense to judge where the water will run and where it is safe to nail the caps. Never place a nail in the exposed metal portion of a valley.

5-59 Overlap the caps so that they will be double-nailed.

5-60 Align and carefully nail the edge of the cap.

5-61 Center the caps at the ridge, overlap each cap by 5 inches and drive one nail on each side about 1 inch from the edge of the cap.

5-62 Continue installing the capping until you reach the rakes.

5-63   Trim the last cap to align with the rakes and drive four nails to secure the corners. Add a dab of roof cement over the four nail heads.

Chapter 6

Angled Roof Sections

Shingling rectangular and square sections of a roof is a matter of selecting a pattern, laying chalk lines, and repeating the shingle pattern until the sections are completed. When you use three-tab asphalt fiberglass-based shingles to cover odd-shaped, angled, or extremely steep roof sections such as hips, wings, dormers, and mansards, more complicated planning is required.

Shingling each of the types of roof sections described in this chapter requires the use of a technique called *filling in* or *backing in*. In brief, filling in a roof section means striking two parallel chalk lines—at a 6-inch width—to square off an angled section of roof so that you will have the longest possible vertical run (the straight pattern) or stepped run (the 45 pattern or the random-spacing pattern) of shingles.

For speed and accuracy, the majority of shingling on angled roof sections should be completed with the work in front of you—right to left if you are right-handed or left to right if you are left-handed. The remaining portion of the roof section is then filled in by working "backwards"; in other words, from the chalk lines toward the rake.

Figures 6-1 through 6-3 show how the application pattern is maintained past an angle in the roof line. Border shingles are installed at the rake, and the shingles are filled in by first high-nailing the courses as they are backed in and then properly nailing them when they are in position.

HIPS

Any of the three patterns described in chapter 5 can be used to shingle a hip roof. The first step is to lay horizontal chalk lines, as described in chapter 4. Because hip roofs have no rakes, carefully determine where you mark the vertical chalk lines.

6-1 Border shingles are installed around a break in the roof section.

6-2 The courses are continued by high-nailing the intersecting shingle.

6-3   The 45 pattern is maintained past the break in the roof line.

If you are working on a new roof or if you have torn off the old shingles, find a rafter at the top left-hand corner of the hip (if you are right-handed) and use it as a reference point for the vertical chalk lines. (See FIG. 6-4.) Mark the vertical lines by following the basic instructions outlined in chapter 4.

Begin shingling the remaining angled portion of the hip sections. As you approach the hip with each course of shingles, use a utility knife or tin snips to trim the shingles at an angle. The last shingle in each course must provide enough coverage so that capping will overlap it.

As you complete the runs, you will require fewer full shingles. Use two-tab, one-tab, and smaller shingle portions to fill in the section. Keep the cutouts aligned. Shingling a hip requires a great deal of cutting and trim work. Be careful!

WINGS

A wing section of a roof can be shingled much like a hip section. First lay horizontal chalk lines, as outlined in chapter 4. Square off the angled section to allow for the longest possible vertical run of shingles. (See FIG. 6-5.)

If you are reroofing over one layer of worn shingles, mark vertical chalk lines 6 inches apart by using the cutouts of the old shingles as guidelines. If you are shingling new work, be extremely careful when you mark off the measurements. Use a rafter at the top of the wing as one reference point and the bottom corner of the wing as the other reference point for the 6-inch-wide vertical chalk lines.

Follow the basic instructions for laying vertical chalk lines given in chapter 4. Make adjustments for the angled rake wing.

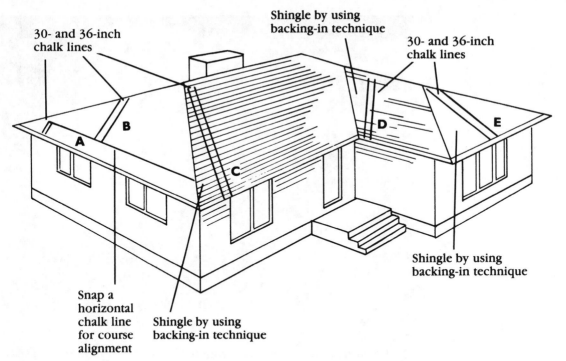

Shingle by using
backing-in technique

30- and 36-inch
chalk lines

30- and 36-inch
chalk lines

Snap a
horizontal
chalk line
for course
alignment

Shingle by using
backing-in technique

Shingle by using
backing-in technique

6-4   Work on a hip roof section by section. In order to keep most of the work in front of you on a hip roof, find a rafter and snap chalk lines so that the bottom third of the hip section can be shingled (A). After the bottom third of the roof has been shingled, snap chalk lines and shingle the top section (B). Square off this hip-roof section by snapping parallel 6-inch-wide chalk lines (C). Square off this section and back the shingles into the valley area (D). Shingle small hip-roof sections by snapping chalk lines at the center of the section (E).

Square off the wing section
with 6-inch-wide, parallel
chalk lines

6-5   A wing section can be squared off for shingling.

Before you begin to nail shingles, make absolutely certain that the measurements and chalk lines are correct. Shingle the squared-off portion of the wing first and then fill in the remaining angled section.

DORMERS

Shingling the tops of dormers generally consists of covering two short sections that taper into valleys. The techniques for installing valleys and shingling valleys are described in chapter 7.

As described in chapter 4, measure for and snap chalk lines. Any of the three shingle-application patterns can be used to cover a dormer, but the straight pattern will probably be the easiest method for you to use.

Begin shingling a dormer at the rakes. Using the straight pattern, you can complete the first runs at the rakes and place a few bundles of shingles within easy reach of the remaining work area.

On one side of the dormer, you will be able to shingle with the work in front of you (FIG. 6-6). On the other side of the dormer, start at the rake

6-6 On the left side of the house, the roof can be shingled from the rake into the valley. On the dormer, back in the shingle pattern from the right rake of the dormer.

and back in the shingles from the rake. Because a dormer usually can be shingled with a few bundles, it is not practical to square off the angled valley section as you would for hip or wing sections.

To properly shingle the main section of a roof that has a dormer, you must carefully plan the shingle pattern so that the shingle courses will match past the top of the dormer and on the other side of the dormer. (See FIG. 6-7.) In order to tie in the courses and keep the shingle cutouts aligned, shingle the rake side of the dormer (where you first began to nail shingles) past the top of the dormer ridge. Also shingle the section of the main roof directly below the dormer and all the way to the other rake.

On the first rake side of the dormer, measure the distance from the bottom course of shingles, including the 1-inch overhang at the border shingle, to the top of the course of shingles immediately above where the dormer ridge line and the valley meet (for example, 30 feet). At the other side of the dormer, scratch a V at the same distance up from the border shingle (for example, 30 feet).

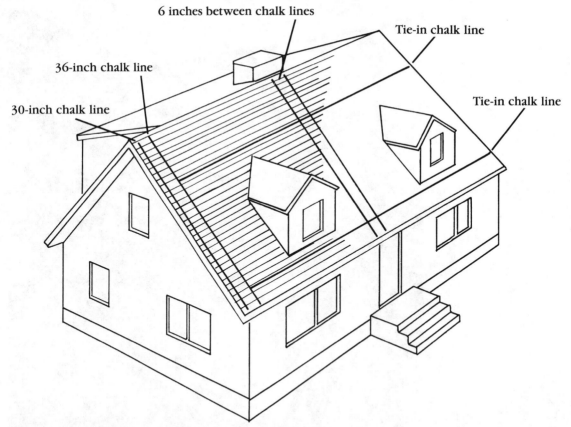

6-7 Using the straight pattern, first shingle up the rake and across the eaves. Once you are past the bottom and the right side of the first dormer, snap chalk lines past the valley section of the dormer. Snap a horizontal tie-in chalk line, high-nail a horizontal tie-in shingle course, and fill in the pattern. Repeat the tie-in technique in order to shingle past the second dormer.

Strike a horizontal chalk line to guide the placement of the tie-in course of shingles. The tops of the tie-in course must touch this line. Position the tie-in course by driving two nails in each shingle no lower than 1 inch from the horizontal chalk line. This technique is called *high nailing*. Later you will nail these shingles in the standard locations above the cutouts and below the adhesive strips. There is no need to remove the high nails.

After you have installed the tie-in course past the dormer, strike two parallel chalk lines 6 inches apart as close to the dormer as possible. The parallel lines must match the cutouts of the tie-in course and the cutouts of the shingles below and to the side of the dormer.

Be extremely careful not to break the proper shingle-pattern sequence. Use the straight pattern to make one run up the parallel lines. Make certain that the application sequence is correct!

Fill in the sections under the tie-in course by lifting the high-nailed shingles and slipping the next course of shingles into position one at a time. Nail the tie-in shingles at the standard locations above the watermarks and below the adhesive strips. Remember to drive four nails into each shingle. Continue backing down the courses until you have finished the section.

MANSARDS

Because you must work from a ladder or a scaffold, a mansard roof can be the most challenging type of surface to shingle for do-it-yourselfers and professionals. If you do not have the proper equipment or the confidence to do a first-rate job—that means safely getting all the shingles on straight in a reasonable amount of time—hire a professional roofer to do this portion of the work.

As with any other roof surface to be shingled, you can use roofing nails or staples to apply felt paper to the deck of a mansard roof. A hammer tacker (FIG. 2-17) will allow you to easily and rapidly staple building paper in place. Use a utility knife to trim felt around windows and at the rakes.

Frequently, mansard roofs are designed with one or more windows. Step flashing or specially designed channel flashing must be installed where the shingles and window meet. Be sure to discuss your mansard-window flashing requirements with a knowledgeable materials supplier.

Begin shingling by applying a course of upside-down border shingles at the bottom edge of the mansard. Allow for the standard 1-inch overhang at the bottom of the facing board or trim board. Do not install drip edge or horizontal border shingles along what would be the "rake" of a less steeply pitched, shingled roof surface.

Because you will be working from a ladder or a scaffold, use the straight pattern to install shingles on a mansard roof. The straight pattern allows you to nail the most shingles per run without having to move the ladder or scaffold. If practical, apply felt over the entire surface of the mansard, strike the 6-inch horizontal chalk line and the chalk line for the bottom border shingle, and then reposition your ladder or scaffolding.

Otherwise, you run the risk of misaligning the shingle cutouts as you snap a series of additional chalk lines up the surface of the mansard.

The professional roofers in FIG. 6-8 are shown using only the factory-made, 6-inch cutouts to align the shingle courses as they install the course across the surface. This technique is not recommended for do-it-your-selfers; too many things can go wrong with this method.

With the first course of felt and the first bottom border shingle in place, install at least one more course of felt up the mansard. Do not install drip edge or horizontal border shingles along what would be the rake of a less steeply pitched, shingled roof surface. Install successive courses of shingles until you reach the bottom of a mansard widow or, if your mansard doesn't have a window, the eaves.

At a window, trim the tops and sides of successive courses just like any other obstacle described in chapter 7. As with a wall or a chimney, flashing around a mansard window must be watertight. Where the top course of mansard shingles meets the eaves shingles, trim the tops of the last course of mansard shingles so that last exposed course of shingles abuts the bottom of the eaves. (See FIG. 6-8J.)

6-8A With a set of pump jacks and scaffolding to provide a stable platform from which to work, shingles can be installed on the deck of a mansard roof much like you would any other roof surface. Adjustable supports are secured at the eaves by spikes at both ends of the roof.

6-8B Pump jacks are designed to be moved as the height of the work changes.

6-8C Felt can be applied using a hammer tacker, felt nails, or roofing nails.

6-8D At the window, each shingle must be positioned, cut, and installed.

6-8E Trim the felt to fit tightly against the window.

6-8F   Because mansards are so visible, each shingle must be carefully aligned.

6-8G   Full shingles can be turned upside down, cut for position, trimmed, and installed.

6-8H   The straight pattern is continued up the mansard roof.

6-8 I   Another course is added to the straight pattern.

6-8J   The last two courses of the mansard shingles are trimmed at the tops of the shingles to fit where the mansard and the eaves meet. The first three courses on the top roof deck have been temporarily left out until the mansard has been shingled.

Chapter 7

Obstacles

*I*nstalling shingles around valleys, chimneys, walls, pipes, and roof vents requires careful planning and careful cutting and trimming of shingles. All of these obstacles are intrusions on roof surfaces; they must be kept watertight.

METAL OR MINERAL-SURFACE VALLEYS

When you measure the lengths of valleys, add 8 inches for each overlap at the ridge and eaves. If you cannot obtain metal or roll roofing in one piece long enough to cover the valley length, add another 6 inches for each overlap between sections of valley material. Install a double layer when you use mineral-surface roll roofing. All valley material must be at least 36 inches wide. (See FIGS. 7-1 through 7-8.)

Before you install the valley material, install drip edge and border shingles at the eaves. For details, see the section on application patterns in chapter 5. Shingle all of the roof sections until just before the runs reach the valley. The bottom section of the valley material must go on top of the border shingles at the eaves on both adjoining sections of the roof at the bottom of the valley.

After the drip edge, border shingles, and first-course shingles are in place, center the bottom section of the valley material where the roof sections are joined. The center of the valley material should extend no more than 1 inch over the bottom of the border shingle. As you position the valley material, several inches of the edge of the valley material will extend over the border shingles. After you have nailed the first section of valley material in place, use tin snips to trim the excess overhang.

Never drive a nail closer than 3 inches from the edge of the valley material. Remember that a great deal of water will be channeled over the valley surface. The easiest way to nail down valley material is to position it, make certain it is centered, and then drive nails every 2 feet or so along

7-1 Galvanized valley material.

7-2 Use tin snips to cut valley material.

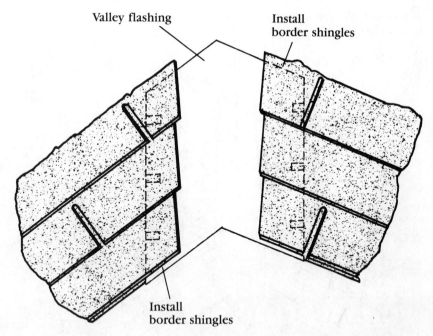

7-3 Border shingles are installed vertically along the valley so that they are aligned with the trimmed shingles.

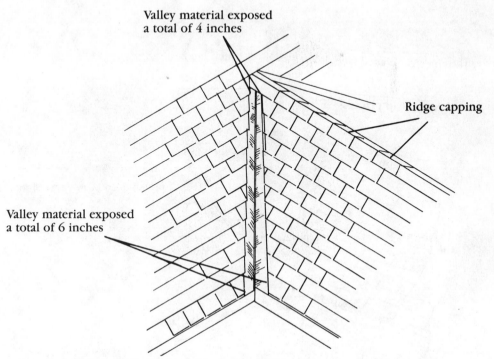

7-4 A valley can be trimmed so that the top is narrower than the bottom.

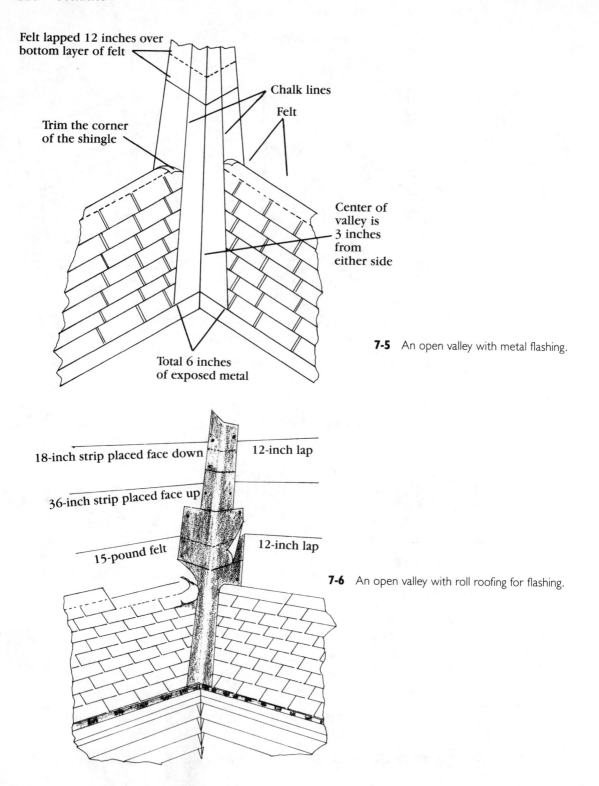

Felt lapped 12 inches over bottom layer of felt

Chalk lines

Felt

Trim the corner of the shingle

Center of valley is 3 inches from either side

Total 6 inches of exposed metal

7-5 An open valley with metal flashing.

18-inch strip placed face down   12-inch lap

36-inch strip placed face up

15-pound felt   12-inch lap

7-6 An open valley with roll roofing for flashing.

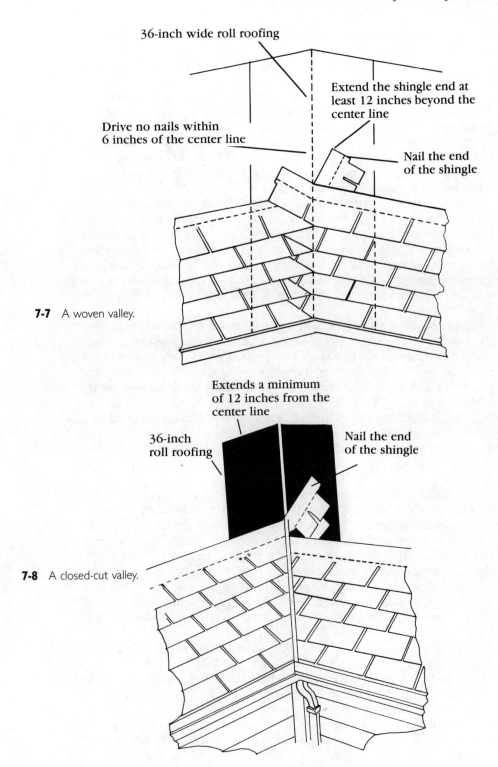

36-inch wide roll roofing

Extend the shingle end at least 12 inches beyond the center line

Drive no nails within 6 inches of the center line

Nail the end of the shingle

7-7 A woven valley.

Extends a minimum of 12 inches from the center line

36-inch roll roofing

Nail the end of the shingle

7-8 A closed-cut valley.

one edge of the metal or mineral-surface roofing. After you have secured one side of the valley section, drive nails on the other side—about every 2 feet—at staggered locations from the other nails, rather than opposite them.

Where sections of valley material overlap, provide at least 6 inches of double coverage. Be absolutely certain to place the top overlapping section above the other valley material section.

At the ridge line, use tin snips to cut the valley material so that it overlaps the intersection of the valley by 6 inches. When you install the other side of the valley, make sure that the valley material again overlaps the ridge lines. Nail carefully at these points.

With the valley material in place, install border shingles along the length of the valley. The border shingles will provide additional protection and help channel the runoff. Border shingles will also provide a straight line on both sides of the center of the valley, which will be quite handy when you trim the scrap from each course of shingles as you roof the valley section. Take your time when you trim the valley shingles. A neat job will look very attractive. An improperly trimmed valley is an eyesore.

To install border shingles along a valley, first find the center of the valley at the eaves. Measure 3 inches from the center on both sides (for a total width of 6 inches). Lightly scratch a set of Vs at two 3-inch measurement points (at the tops and bottoms of the border shingles). Position full shingles with the granular side up and the cutouts facing away from the center of the valley.

The shingle tops must be aligned with the measurement marks. Check to determine if you have a full 6 inches of exposed valley between the two border shingles. Drive two or three nails in the tabs of each shingle to hold them in place. Do not drive nails near the exposed valley material.

At the ridge line, install two more border shingles. Snap chalk lines at the tab side of the border shingles (away from the middle of the valley) to serve as guidelines, and fill in the valley border with shingles.

You can now resume roofing the sections by following the application pattern. When you are shingling at the bottom of a dormer valley, check to make certain that the border shingle and the first-course shingle directly below the dormer valley are both under the valley material. Install the next shingle course and trim the scrap along the valley border. Use tin snips to trim the scrap from each course as you nail the course in place.

In many cases, you will find it convenient to position the shingle, cut it, and then nail it in place. At other times, you will find it necessary to extend a full shingle into the valley (where the end will be trimmed) by first inserting a one- or two-tab shingle. In this way you can avoid driving a nail into the valley material.

Be extremely careful not to cut into the valley material. Any cut or torn, exposed valley sections must be replaced with watertight valley material.

CHIMNEYS

Old-Work Rake Chimneys A chimney that is constructed along a rake must be flashed at three sides where it meets the roof. Because the intrusion is at the rake, the shingle application pattern should be easy to continue. If you are reroofing over one layer of worn shingles, do not attempt to lift, remove, or replace the chimney flashing unless it is in poor shape.

For reroofing jobs, trim the shingles so that they fit snugly against the flashing, avoid driving nails into the flashing, then apply a generous amount of roofing cement around the chimney flashing. Use the most cement at the back and at the side of the chimney where runoff will pass over the flashing.

New-Work Rake Chimneys If you have torn off the old shingles or if you are shingling new work and the chimney is at the rake, you must flash and counterflash the chimney. Where you tear away shingles from a chimney, try to carefully bend up the counterflashing—without removing it—so that you can use it again. If you must install new counterflashing, use 1-inch roofing nails to attach the counterflashing at the mortar joints.

Nail several runs of shingles until you reach the course just below the bottom of the chimney. As you add courses around the chimney, the tops and sides of the shingles must be trimmed to fit. In addition, you must maintain the alignment of the cutouts, and the rake side of the shingles must be trimmed even with the border shingles. You will find it convenient to first position the shingle, then cut it and nail it.

As you nail shingles at the corners and sides of a newly constructed chimney, you must install flashing. Purchase 5- × -7-inch aluminum step flashing or cut your own flashing. (See FIGS. 7-9 through 7-11.)

Allow a 2-inch overlap with each piece and install the 7-inch portion of the step flashing against the roof surface. First position a shingle, nail a piece of step flashing in place (nail only on the 7-inch portion, using two nails at most), position the next shingle, and then install another piece of step flashing. Continue weaving the flashing and shingles. Remember that each piece of step flashing must overlap the one below it by 2 inches.

When you reach the shingle course that is even with the bottom of the chimney, nail the first shingle in place. A row of overlapped step flashing or one continuous piece of flashing now must be installed flush against the bottom of the chimney.

At the bottom and top corners of the portion of the chimney adjacent to the rake, you must form watertight "boxes" of flashing. Use two pieces of step flashing to form a box. Use tin snips to cut one piece of flashing halfway into the crease. Do not detach the cut section. Bend the cut section so that, when it is installed, it will wrap around the corner of the chimney. Cut the second piece of flashing so that it will bend in the opposite direction. When it is installed, the box must have the higher piece on top so that water will flow over the flashing.

Install the next shingle flush with the side of the chimney. Maintain the pattern of shingle cutouts. As you install each piece of step flashing, make certain that the flashing is overlapped. The flashing should not

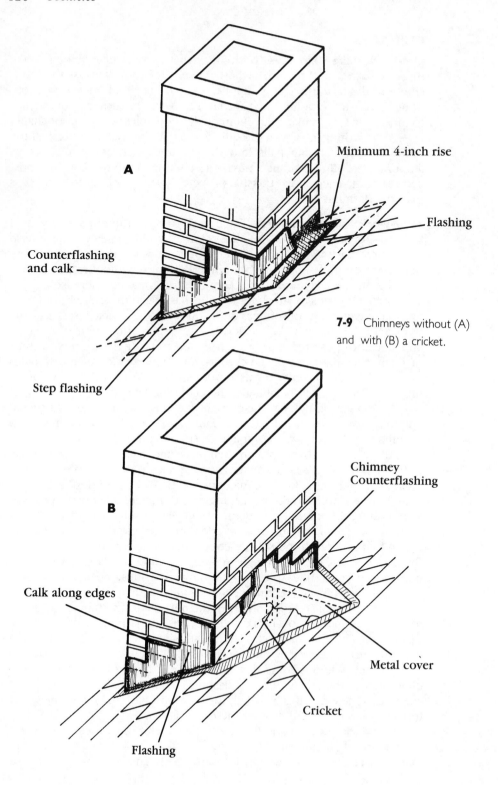

Minimum 4-inch rise

Flashing

A

Counterflashing
and calk

7-9 Chimneys without (A)
and with (B) a cricket.

Step flashing

Chimney
Counterflashing

B

Calk along edges

Metal cover

Cricket

Flashing

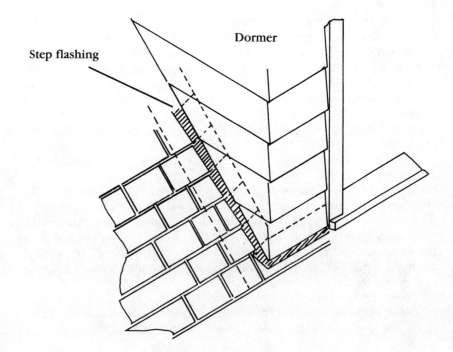

7-10 Flashing at a dormer wall.

7-11 Application of flashing at the front of a chimney.

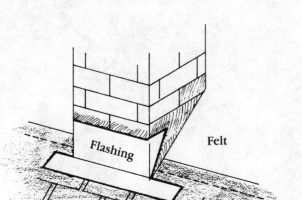

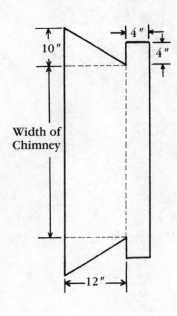

extend past the bottoms of the shingles; in other words, there should be no metal showing except along the chimney side.

When you reach the top corner of the chimney, cut off the top section (all but the tabs) from one or two shingles, depending upon the width of the chimney, and nail in place a strip that extends about 3 inches past the chimney corner. Trim the other side even with the rake border shingles.

Install flashing on top of the shingle strip along the back of the chimney, and nail the second corner flashing box in place. Trim the next few courses of shingles to fit snugly against the chimney and continue the application pattern.

Middle-of-the-Roof Chimneys   In order to successfully shingle around a chimney that intrudes upon the center of a roof section, you must plan the shingle pattern so that the courses will meet past the top of the chimney. In addition, you will have to flash the chimney—this time on four sides—as described in the preceding section. Step flashing can be used around the chimney, or you can waterproof the chimney with flashing cut from galvanized valley material.

Read the section on dormers in chapter 6. Use the same basic techniques to tie in the shingles. Shingle the section directly below the chimney, strike chalk lines, and install a tie-in course. (See FIGS. 7-12 through 7-47.)

7-12   Tearing out worn chimney flashing with a pry bar.

7-13 (Above) The old flashing is torn away from the back of the chimney.

7-14 (Left) A claw hammer makes the work easier.

7-15 (Above) Flashing can be cut from a roll of galvanized metal. Use a chalk line to mark an even line.

7-16 (Left) Use tin snips to cut the metal along the chalk line.

7-17   Bend the side flashing.

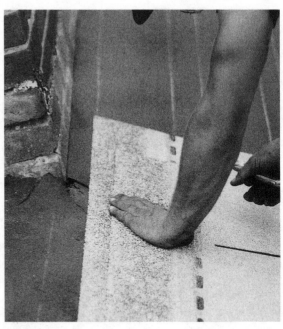

7-18   Trim the shingle courses to fit tight against the bottom of the chimney.

7-19   The shingle pattern is maintained.

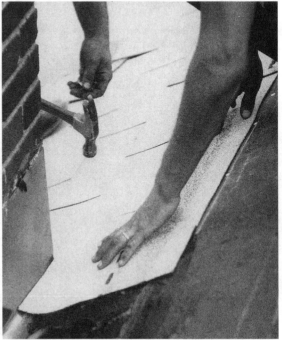

7-20   Roofing cement will provide additional protection at the back of a chimney.

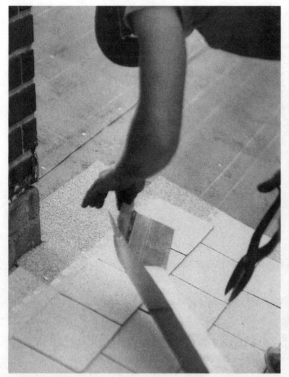

7-21 (Above left) At the back of the chimney, position the new flashing.

7-22 (Above right) Carefully drive roofing nails into the flashing at the mortar joints.

7-23 (Left) Carefully position the flashing at the chimney corner.

7-24 Position the side chimney flashing. The top of the flashing can be cut to give the appearance of steps.

7-25 The flashing is placed on top of the shingle course at the chimney corner.

7-26 At the chimney corner, one shingle is placed under the flashing, and the flashing is attached with roofing nails driven into the mortar joints.

7-27 (Left) Trim the shingles so that they fit close to the flashing.

7-28 (Below) Continue the shingle pattern.

7-29 (Left) Cut on the back of the shingle so you can carefully size the tab.

7-30 (Below) The tie-in course must be carefully planned.

7-31 The straight pattern allows you to quickly shingle past the chimney.

7-32 Once the shingle pattern is tied in, the area next to the chimney can be filled in.

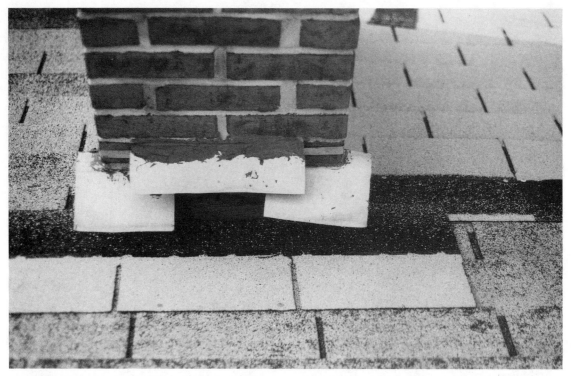

7-33 If the chimney flashing is not removed, place a border shingle behind the chimney to provide flashing.

7-34 Maintaining the 45 pattern past the chimney.

7-35 (Above) Cut the tabs to fit snugly against the chimney side.

7-36 (Left) Maintain the pattern with a cut shingle.

7-37 Install a two-tab shingle so that another two-tab piece can be cut to fit against the side of the chimney.

7-38 If the watermark is closer to the chimney than 6 inches, trim a two-tab piece to fit. Don't attempt to nail a very small piece of shingle next to the chimney.

7-39 Add courses to maintain the 45 pattern along the side of the chimney.

7-40 Install tabs to maintain the pattern.

7-41   Trim the corner shingle to fit.

7-42   The shingle pattern is continued past the chimney.

7-43   Position the tab before nailing.

7-44   Cut the tab to fit the corner.

The chimney must be watertight, so carefully install the flashing and avoid driving nails near the edges of the chimney. Apply plenty of roofing cement to all sides of the chimney and caulk the flashing seams. (See FIGS. 7-48 through 7-50.)

7-45 Install the corner shingle and continue the pattern.

7-46 Fill in the courses by lifting tabs.

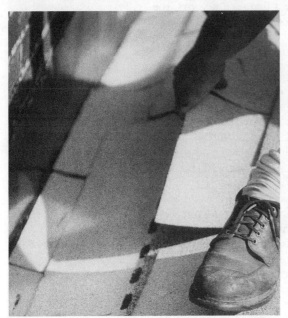

7-47 Courses above the tie-in shingles have been high-nailed.

7-48 Apply roofing cement along the chimney sides.

7-49 Apply plenty of roofing cement at the back of the chimney.

7-50 Caulk can be applied to flashing seams.

VENT FLANGES AND PIPE COLLARS

Shingles around vents and pipes can be installed without chalk lines and elaborate tie-in courses. A vent and pipe will have a flange. On vents, the flange probably will be part of the unit. Pipes are flashed with a metal or polyurethane collar; the flange will be part of the collar (FIG. 7-51). Shingle the roof section until you have at least one full course below the obstacle. (See FIGS. 7-52 and 7-53.)

▶ **Application Tip**   Avoid two common errors when you are installing shingles around vents and pipes. First, note that most pipe collars have a tapered neck. The tapered portion faces down the roof when the collar is properly installed. Second, at least one course of shingles directly below the pipe must go under the flange. At times, it will be necessary to place a portion of more than one course under the flange. A substantial leak will result from the improper installation of a flange.

Trim circular cuts in one-tab shingles around a pipe. Where possible, push the tab into position, make scratch marks at the appropriate cut locations, and then use a utility knife or tin snips to make partial cuts. Again place the tab in position and determine how much more you need to cut. The shingles should fit as close as possible to the pipe and collar. (See FIGS. 7-54 through 7-58.)

Apply a generous amount of roofing cement where the collar and the shingles meet and over any exposed nail heads.

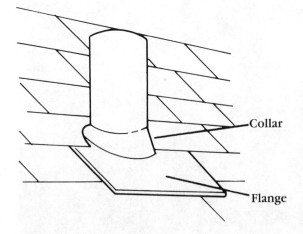

7-51 A flange must be installed on top of at least one shingle. The amount of exposed flange will depend upon where the shingle course intersects the pipe.

7-52 An old flange can be used with new shingles.

7-53 Install the flange over the shingle course.

7-54 Position and mark the shingle for cutting.

7-55 Trim the shingle around the pipe flange.

7-56 Trim the next shingle.

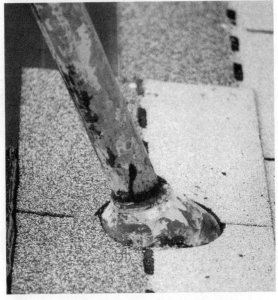

7-57 Shingles should fit around, but not over, the flange.

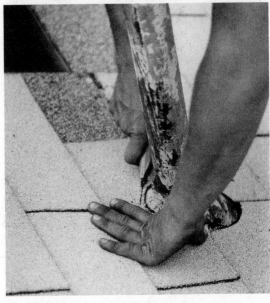

7-58 Continue the pattern past the pipe.

WALLS

If you are shingling over one layer of worn shingles, do not lift or remove old step flashing from alongside a wall unless the flashing is badly worn. As you install the shingles, trim them as close to the wall as possible.

On new work, install 5- × -7-inch step flashing with the 7-inch portions flat against the roof. (See FIGS. 7-59 and 7-60.) At the bottom corner where the wall and the roof meet, install box flashing as you would at the corner of a chimney. Weave the shingles and step flashing as you make the run. (See FIGS. 7-61 through 7-68.)

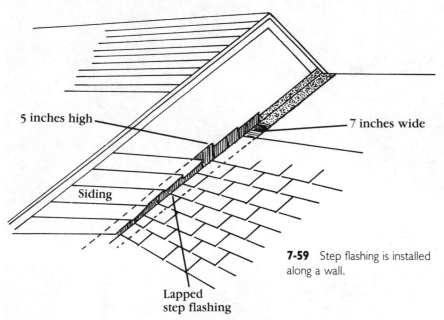

7-59 Step flashing is installed along a wall.

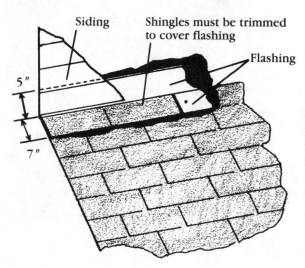

7-60 Step flashing can be installed against a vertical wall to ensure a watertight seal.

At the ridge line, use tin snips to cut a piece of step flashing halfway along the crease. Cut another piece of step flashing along the opposite side of the crease. Fold the edges down, slip the pieces together, and form a flashing "tent" at the ridge line. Nail down the tent after you have shingled and flashed both sides of the roof section.

7-61   Install a border shingle where the rake and wall meet.

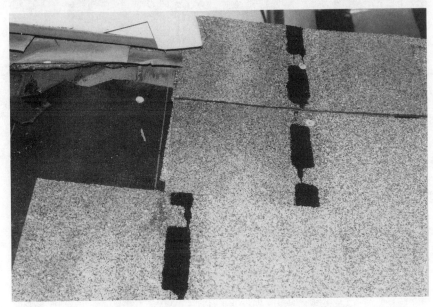

7-62   Continue the shingle pattern by installing tabs.

7-63 Position a full shingle, mark it where it is to be cut, and then trim it to fit.

7-64 The trimmed shingle (Fig. 7-63) must be installed under the first piece of step flashing. On this reroofed section, the step flashing was not completely removed from along the wall. Nails holding the flashing in place were removed so that the shingle courses could be weaved with the flashing.

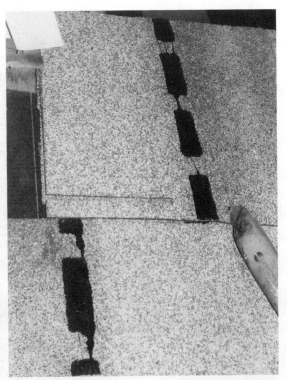

7-65 By turning the shingle upside down, you can easily find the proper place to cut.

7-66 Again, turn the shingle upside down in order to mark the remaining portion of shingle to be trimmed.

7-67 With the shingle properly trimmed to fit against the wall, the second piece of step flashing is placed over the shingle and nailed about $1/4$ inch from the edge. Don't nail the step flashing too close to the wall. Compare the position of the flashing with that shown in Figs. 7-64 and 7-68.

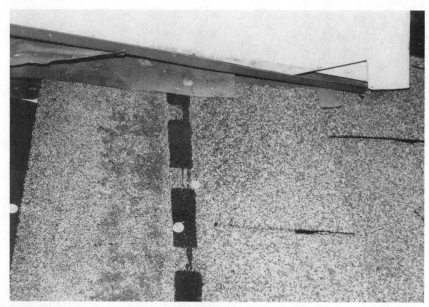

7-68 Continue weaving step flashing and shingles. The flashing should be even with or slightly beyond the watermarks.

Chapter *8*

Roll Roofing

*R*oll roofing is appropriate for use on low-slope roofs such as sheds, porches, and garages. It is best when used on surfaces with low visibility where appearance is not important. Roll roofing can be applied horizontally or vertically, but vertical installation will not be as watertight because the very long vertical seams are subject to seepage. The advantages of using roll roofing include low cost, ease of application, and suitability for surfaces with as gentle a pitch as 1 inch per foot. E-Z Roof roll roofing has the added advantage of a peel-and-stick self-adhesive backing that provides a quick bond without the need for spreading gallons of roofing cement.

On the other hand, roll roofing does not have aesthetic appeal (it comes in 36-foot-long, 3-foot-wide strips), and it will wear out in about half the time as most shingles. Roll roofing is manufactured with the same basic materials as asphalt shingles, but its weight per 100 square feet is 90 to 180 pounds. Asphalt, fiberglass-based shingles range in weight from 210 to 325 pounds per 100 square feet. The difference in weight is the main reason for roll roofing's short life span.

SINGLE-COVERAGE APPLICATION STEPS

Prepare the roof surface as described in the sections on tearing off worn materials, laying felt, and installing drip edge in chapter 4. You will need the following tools and equipment.

- Stiff-bristle broom.
- Ladder.
- Nail bar.
- Nail pouch.
- Roofing cement.
- Roofer's hatchet.

- Roofing nails (1-inch galvanized or aluminum).
- Tape measure.
- Tin snips.
- Trowel.
- Utility knife.

First Course   With the felt and drip edge in place, apply a horizontal starter course the entire length of the eaves. If the starter course is longer than 36 feet, overlap the joint by 6 inches. The starter course should extend over the rakes and the eaves by 1 inch. (See FIG. 8-1.)

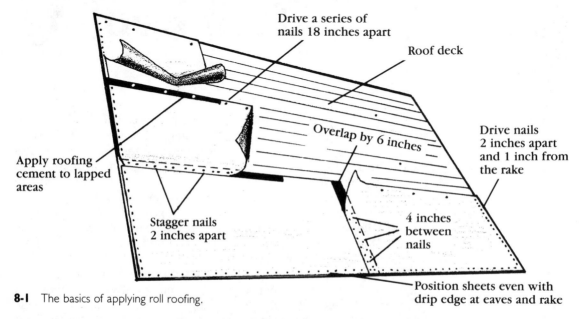

Drive a series of
nails 18 inches apart

Roof deck

Apply roofing
cement to lapped
areas

Overlap by 6 inches

Drive nails
2 inches apart
and 1 inch from
the rake

Stagger nails
2 inches apart

4 inches
between
nails

Position sheets even with
drip edge at eaves and rake

8-1   The basics of applying roll roofing.

Along the top of the starter strip, about $1/2$ inch from the top, nail the course every $1^1/2$ feet. At the bottom and at the rakes, drive nails 2 inches apart and about 1 inch from the edges.

Second Course   Position the second course so that it overlaps the top of the starter course by at least 2 inches. Most manufacturers of roll roofing include painted guidelines on their products for easy alignment of courses.

If the roof surface is longer than 36 feet, do not align the joints or leaks will result. Where joints are necessary, use a trowel or a stiff-bristle brush to apply a smooth coat of roofing cement, and then embed the overlapping sheets. Nail the lap joint at 2-inch intervals.

Third to Final Courses   Continue applying courses until you reach the ridge or a wall. If the roof has a ridge, trim the top course even with the ridge on one side. Lap the last course from the other side of the roof and nail it snugly against the deck.

Using a chalk line and tin snips or a utility knife, make ridge capping by cutting a strip of roll roofing in half lengthwise. When you position the capping course, make certain the capping strip is straight and without bulges or wrinkles.

If the roll-roofing courses must abut a wall, cut L-shaped galvanized step flashing, about 42 inches long, for each course. Weave the roofing and the metal so that the flashing method resembles the techniques used to waterproof the sides of a chimney, as described in chapter 7.

DOUBLE-COVERAGE APPLICATION STEPS

In order to allow for double coverage with roll roofing, manufacturers produce *selvage-edge* roll roofing that is intended to provide 17 inches of mineral-surface roofing and 19 inches of saturated felt in one 36-inch-wide, 36-foot-long sheet. Some manufacturers of selvage edge require the use of cold, asphalt-based roofing cement with their products. Other manufacturers require the use of hot asphalt. The use of cold, asphalt-based roofing cement makes the job much easier for the do-it-yourselfer.

Horizontally applied selvage-edge roll roofing can be used on a roof deck with a rise of 1 inch or more per foot. Prepare the roof surface as described in the sections on tearing off worn materials, laying felt, and installing drip edge in chapter 4. In addition to the tools and equipment listed at the beginning of this chapter, you will need a small, stiff-bristled broom and the type and quantity of roofing cement recommended by the manufacturer of the selvage-edge roll roofing you choose.

First Course With the felt and drip edge in place, make a horizontal starter course by cutting the selvage portion of a strip long enough to reach the entire length of the eaves. Save the mineral-surface portion of the roll for the last course. If the starter course must be longer than 36 feet, overlap the joints by 6 inches. The starter course should overlap the rakes and the eaves by 1 inch.

Secure the selvage starter course by driving roofing nails at 6-inch intervals along the rake, along the eaves, and staggered across the center.

To complete the first course, roll out a 36-inch-wide strip of selvage-edge roll roofing long enough to cover the starter course. Align the material so that the mineral-surface (granule) portion of the strip exactly matches the coverage of the starter strip. Nail only the selvage portion in place. Do not drive any nails in the mineral-surface portion of the roll roofing.

When the course is in position, very carefully lift up the mineral-surface portion and gently flip it over to expose the selvage starter course. Use a short-bristle broom to apply roofing cement, as specified by the manufacturer, to the selvage starter course. Gently flip the mineral-surface portion of the roll roofing back into position. Be careful not to tear the strip. (See FIGS. 8-2 and 8-3.)

Second Course Apply the next and succeeding course by lapping the previous selvage-edge layer with the mineral surface of the next strip of roll roofing. Where joints are necessary, overlap the edges by 6 inches and apply roofing cement along the joint. Never align joints in successive courses or a leak will develop.

Last Course   When you are ready to apply the last course of mineral-surface material, retrieve and nail in place the strip left over from the selvage-edge starter course. When all of the courses have been applied, use your feet to ensure that all of the sheets of roll roofing have good adhesion.

When you leave the roof, your shoes, tools, clothing, arms, and hands will almost certainly have a good amount of roofing cement on them. It's just about unavoidable when you do the job properly.

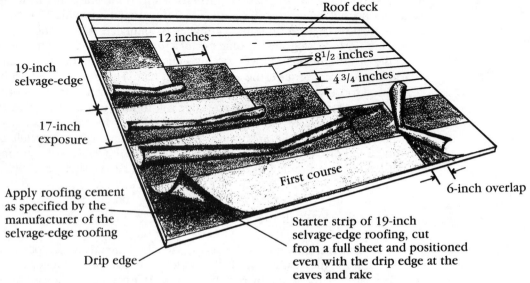

8-2 Installing double-coverage roll roofing.

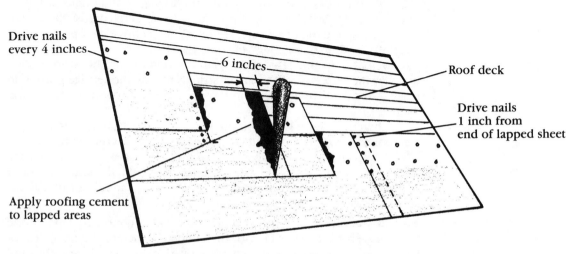

8-3 Applying roofing cement to the lapped areas.

Chapter **9**

Wood Shingles and Shakes

Redwood or cedar wood shingles or hand-split shakes make attractive alternatives to asphalt fiberglass-based shingles. Compared to fiberglass shingles, however, material costs and shingling time required per square are much higher for wood shingles or shakes. Before you decide to use wood shingles or shakes, make certain your local building code permits the installation of wood shingles or shakes on roofs.

Wood shingles are produced in uniform 16-, 18-, and 24-inch widths. Wood shakes are split to expose at least one natural-grain textured surface, and therefore, will present an irregular pattern when installed. Packed in bundles that are easy to handle, wood shingles and shakes are applied using the basic principle of roofing: overlapped layers of water-shedding materials.

Installation techniques for shakes and shingles are similar. Exceptions are that an 18-inch-wide interlay strip of 15-pound felt must be applied beneath each course of shakes, and the amount of exposure to the weather for shakes is greater than for shingles. While shakes or shingles can be applied to a deck without plywood (spaced sheathing) or standard sheathing, or over one layer of worn asphalt shingles or roll roofing, wood shingles must be installed over furring strips to provide the proper ventilation that will ensure long shingle life.

Valley flashing should extend 10 or more inches on either side of the center of the valley. Walls, vents, and chimneys are potential trouble spots. See chapter 7 for descriptions of how to deal with these common roof obstacles. Be especially careful when you flash pipes. Remember that the top of the flange must be covered by the roofing material and the bottom of the flange must go over the top of one or two courses of the roofing material.

TOOLS AND EQUIPMENT

Using the right tools and equipment for the job will make the work go easier and faster. In addition to the items described in chapter 2, you should use a 4-foot-long straightedge tacking board (FIG. 9-1) to align the courses as you work across a roof section. A wooden roof seat will be useful on a roof with a not-too-steep pitch, and a hand-held power saw will be convenient for angling shingles or shakes.

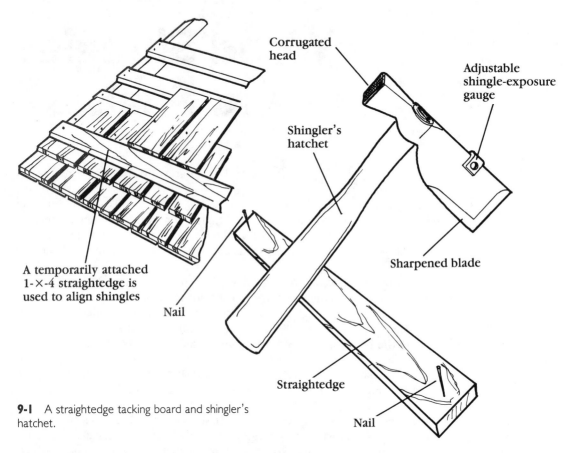

Corrugated head

Adjustable shingle-exposure gauge

Shingler's hatchet

Sharpened blade

A temporarily attached 1-×-4 straightedge is used to align shingles

Nail

Straightedge

Nail

9-1 A straightedge tacking board and shingler's hatchet.

Hatchet Gauge The adjustable gauge on a shingling hatchet will provide you with a convenient way to check for proper shingle exposure. The gauge can be set, for example, at $7^1/2$ inches from the head of the hatchet (FIG. 9-2). With a shingle in position, you can use your hatchet to gauge whether a shingle is ready to be nailed in place. (See FIG. 9-3.)

Roofer's Seat A roofer's seat (FIG. 9-4) can only be used to install wood shingles or shakes. The seat would badly mar fiberglass shingles and it cannot be used on a metal roof.

You can make your own seat by cutting an 18-inch-long base (seat) and a 15-inch-long support base from 1-×-12 lumber. You will also need strips of wood for the "gripper" nails.

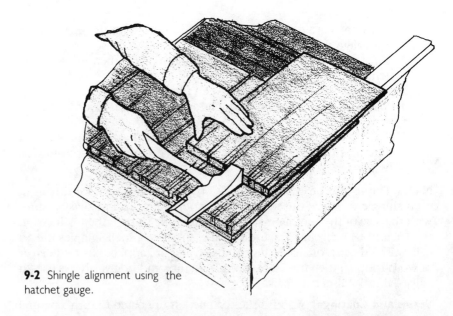

9-2 Shingle alignment using the hatchet gauge.

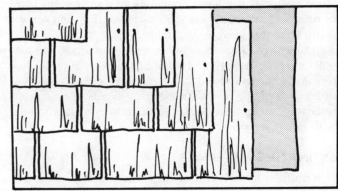

9-3 Proper shingle spacing.

9-4 A roof seat will help you install wood shakes or shingles.

1. Place the base on the roof facing the ridge. With the seat side on top, align the front of the seat and the support base.
2. Lift the back of the seat so that it is level, and then measure the length of boards needed to fit the sides.
3. Cut and install the pieces between the seat and base.
4. Cut several 3-inch-wide, 10-inch-long strips of plywood and drive several 1-inch roofing nails through each strip.
5. Nail the strips to the base of the slat with the nail points facing out. The nails will grip the roof and keep the seat in position as you work.

Nails   The length of galvanized nails—two and only two are driven in each shingle or shake—used to secure wood shingles and shakes varies with the size of the roofing material and the type of base they will cover. For 16- and 18-inch shingles, use 3d nails, and for 24-inch shingles use 4d nails. Use 6d nails for shakes unless the shakes are applied over one layer of worn roofing; then use 7d or 8d nails. At hips and ridges, always use 8d nails. Table 3-2 lists nail sizes.

Stains and Coatings   Wood shakes can be left uncoated to turn a natural attractive gray or you can apply one of a variety of clear coatings, stains, or paints. If the shakes are left untreated, they can sometimes develop an unattractive or uneven appearance.

A clear coating will provide some protection against wood rot and it will help repel moisture. A semitransparent stain will allow the wood texture and grain to show, and provide a more Colonial look for the roof. In order to apply paint, untreated cedar needs one or more coats of bleed-resistant primer. Follow the manufacturer's instructions in accordance with label directions.

ESTIMATING MATERIALS

Determining the amount of materials you will need to roof your home with wood shingles or shakes is a little more complicated than figuring for fiberglass shingles. Because the amount of shingle or shake exposure will vary with the pitch of your roof, you must first determine the roof's total square footage, and then add a percentage to compensate for the pitch (TABLE 9-1).

Include in your estimate total 1 square of shingles for every 100 linear feet of hips and valleys. For starter shingles, one square will cover about 240 linear feet. Allow two squares of shakes for every 120 linear feet of hips, valleys, and starter course. It takes four bundles to make a square (100 square feet) of shingles. There are five bundles per square of shakes.

SHINGLE APPLICATION

To apply wood shingles over a layer of worn wood or asphalt shingles, first cut back about 6 inches of the old roofing along the eaves and rakes to provide a neat appearance. Nail strips of lumber along the rakes and eaves so that all of the roof-section surfaces will again be level. If you have

Table 9-1 Recommended Exposure for Wood Shingles and Shakes.

Do not use wood shingles on a roof with less than 3″ in 12″ pitch			
Shingle length	*Shingle thickness*	*Slope: 3 in 12*	*Slope: 4 in 12*
16″	5 butts in 2 inches	$3^3/4$″ exposure	5″ exposure
18″	5 butts in $2^1/4$ inches	$4^1/4$″ exposure	$5^1/2$″ exposure
24″	4 butts in 2 inches	$5^3/4$″ exposure	$7^1/2$″ exposure
Shake length			
18″		not recommended	$7^1/2$″ exposure
24″		not recommended	10″ exposure
Length	*Exposure*	*Coverage*	
16″ Shingle	$7^1/2$″	150 square feet	
18″ Shingle	$8^1/2$″	154 square feet	
24″ Shingle	$11^1/2$″	153 square feet	
18″ Shake	$8^1/2$″	85 square feet	
24″ Shake	$11^1/2$″	115 square feet	

decided to use drip edge, install it now along the eaves and rakes, as described in chapter 4.

Ventilation Strips   If the shingle exposure is $5^1/2$ inches or more, use 1 × 3s for ventilation furring strips. Use 1 × 2s if the shingle exposure is less than $5^1/2$ inches. Nail the furring strips parallel with the eaves and spaced at equal distances so the shingle exposure remains the same for every course. (See FIGS. 9-5 through 9-7.)

Starter Course   At both rakes of the roof section, position a shingle with the butt overhanging the eaves by $1^1/2$ inches. The rakes should have a 1-inch overhang. To secure the first end shingles, drive nails about $1^1/2$ inches above the exposure line. Install the starter course across the eaves by spacing the shingles $1/4$ inch apart. The starter course must be doubled, and it can be tripled if you prefer a more textured appearance at the eaves.

Offset all vertical joints between the shingles by $1^1/2$ inches. The joints permit expansion of the wood during hot weather and thereby prevent buckling of the wood.

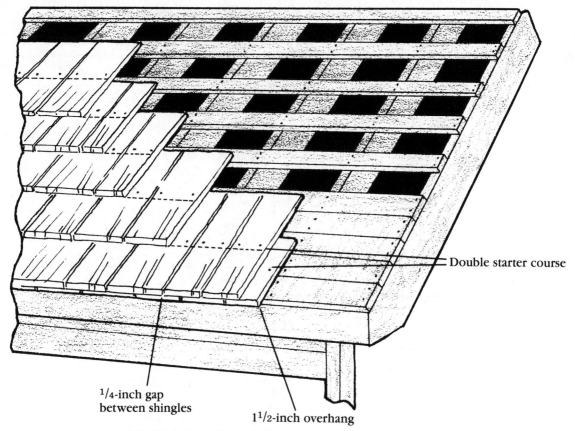

Double starter course

$^1/_4$-inch gap
between shingles

$1^1/_2$-inch overhang

9-5 Ventilation strips are installed when wood shingles are applied.

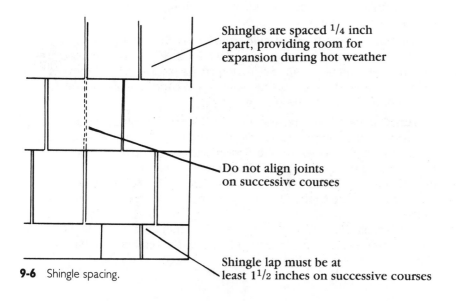

Shingles are spaced $^1/_4$ inch
apart, providing room for
expansion during hot weather

Do not align joints
on successive courses

9-6 Shingle spacing.

Shingle lap must be at
least $1^1/_2$ inches on successive courses

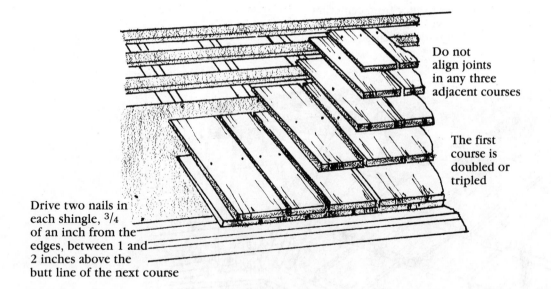

Do not
align joints
in any three
adjacent courses

The first
course is
doubled or
tripled

Drive two nails in
each shingle, $3/4$
of an inch from the
edges, between 1 and
2 inches above the
butt line of the next course

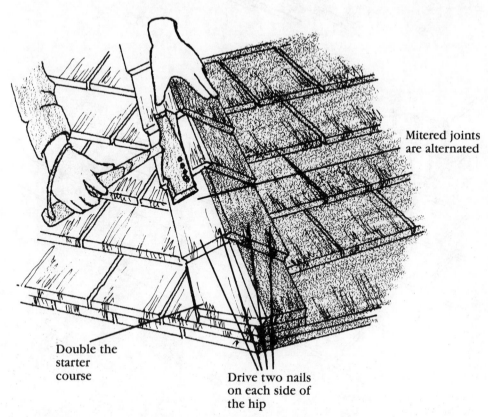

Mitered joints
are alternated

Double the
starter
course

Drive two nails
on each side of
the hip

9-7 Wood shingle application techniques. Continued on p. 156.

**Width of valley from centerline
varies with the pitch of the roof:
7 inches to 10 inches**

9-7 Continued.

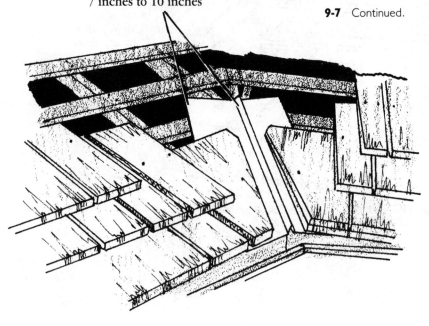

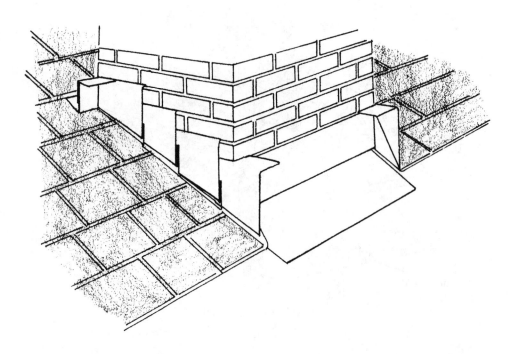

Subsequent Courses With each subsequent course, no joints in any three adjacent courses should align. For hips and ridges, use special factory-assembled units. Saw shingles that extend into valleys so that the shingles fit the angle of the valley center. Do not install shingles parallel with the angle of a valley.

SHAKE APPLICATION

To apply wood shakes over a layer of worn shingles, first cut back about 6 inches of the old roofing material along the eaves and rakes to ensure a neat appearance. Nail strips of lumber along the rakes and eaves so that all of the roof-section surfaces will again be level. If you decide to use drip edge, install it now along the eaves and rakes, as described in chapter 4.

Starter Course Roll out and install a 36-inch-wide course of felt the length of the eaves and trimmed at the rakes. Either 15- or 18-inch shakes can be used for the underlying starter course.

At both rakes of the roof section, position a shingle with the butt overhanging the eaves by 2 inches. The rakes should have a $1^1/2$-inch overhang. To secure the first end shingles, drive nails about $1^1/2$ inches above the exposure line. Install the starter course across the eaves by spacing the shingles $1/2$ inch apart. The starter course must be doubled, and it can be tripled if you prefer a more textured appearance at the eaves. (See FIGS. 9-8 through 9-10.)

Offset the vertical joints between the shakes by $1^1/2$ inches. The joints permit expansion of the wood during hot weather and thereby prevent buckling of the wood.

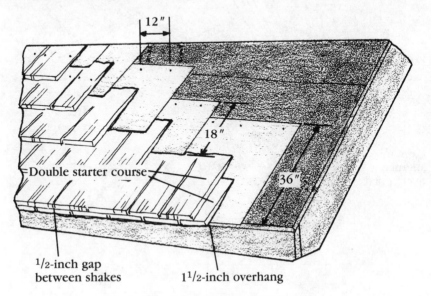

9-8 Wood shake application.

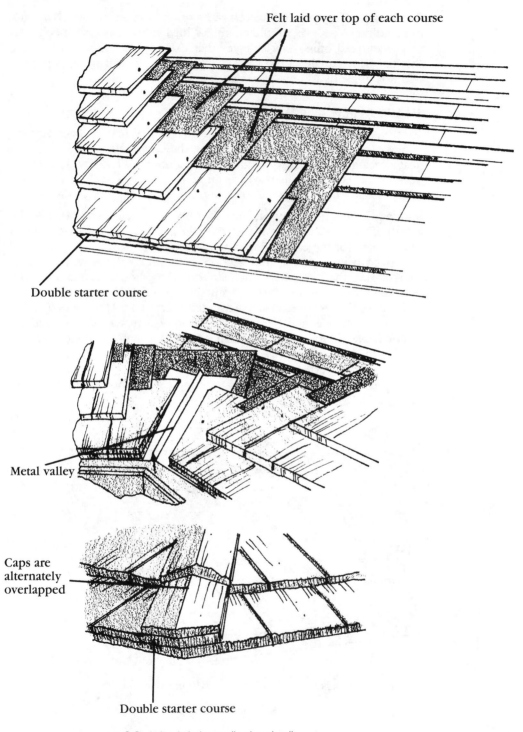

Felt laid over top of each course

Double starter course

Metal valley

Caps are
alternately
overlapped

Double starter course

9-9 Wood shake application details.

Shake Application   159

Subsequent Courses   Cut an 18-inch strip of felt the length of the starter course and install it across the top of the shakes. The bottom of the 18-inch felt course should lap the shakes 20 inches above the butt line of the shakes. Using a straightedge tacking board or the gauge on your roofing hatchet to ensure the proper shake exposure, install a course of shakes over the felt.

You can save time by laying three or four more courses of 18-inch-wide felt strips on the roof section, and then tucking the shakes under the felt as you apply each course. Don't apply felt to an area larger than where you will be able to install shakes during the same day. Use 1-inch roofing nails to secure only the top edge of the felt. If you nail the felt at any other points, you will not be able to install the shakes properly.

Be extremely careful when you apply the felt strips. Never walk on felt that has not been securely nailed.

Continue interlaying felt with shake courses as you roof the building. No joints in any three adjacent courses should align. For hips and ridges, use special factory-assembled units. Saw shingles that extend into valleys so that the shingles fit the angle of the valley center. Do not install shingles parallel with the angle of a valley.

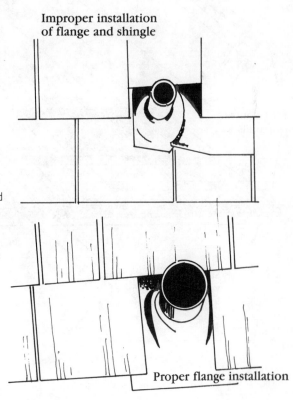

Improper installation
of flange and shingle

9-10 Vent flanges must be installed under at least one course.

Proper flange installation

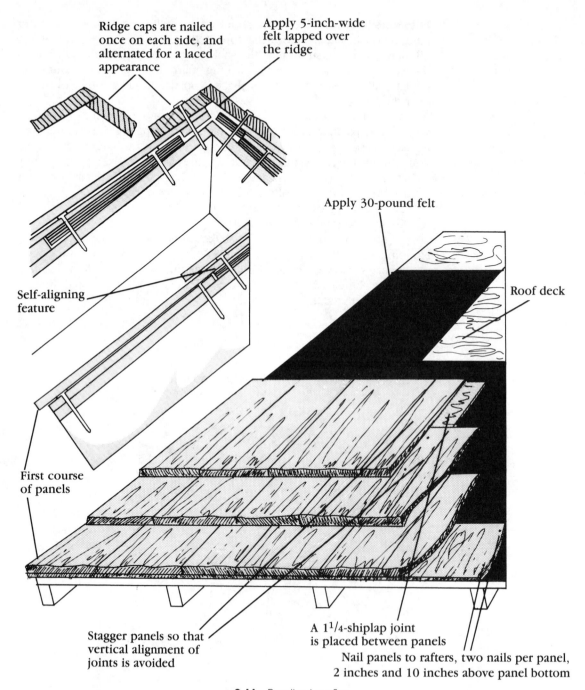

Ridge caps are nailed once on each side, and alternated for a laced appearance

Apply 5-inch-wide felt lapped over the ridge

Apply 30-pound felt

Roof deck

Self-aligning feature

First course of panels

Stagger panels so that vertical alignment of joints is avoided

A $1^{1}/_{4}$-shiplap joint is placed between panels

Nail panels to rafters, two nails per panel, 2 inches and 10 inches above panel bottom

9-11 Panelized roofing.

PANELIZED ROOFING

Installing individual shakes or shingles can take three to five times as long as installing fiberglass shingles. One way to reduce the time needed to complete the job is to use 8-foot-long panels, such as those manufactured by Shakertown Corporation.

The cedar shingles and shakes are bonded by waterproof adhesives to form three-ply panels. The panels do not need sheathing and can be nailed directly to studs. Shakertown two-ply panels can be applied to sheathing or to furring strips.

A self-aligning feature and the need for only two nails per panel per rafter speeds the work. No straightedge or gauges are needed to keep the courses aligned. (See FIG. 9-11.) Complete application instructions are included with each bundle of panels.

Masonite Corporation makes a fiberboard-panel roofing product, from highly compressed wood fibers, called Woodruf. The 12-×-48-inch panels, that are 50 percent denser than natural wood, look convincingly like wood shakes when they are installed on a roof. You will need 32 to 36 Woodruf panels to cover 100 square feet. In comparison, it will take about 200 wood shakes or 80 asphalt shingles to cover the same 100 square feet of roof surface. If left uncoated, the panels will weather to a natural gray cedar color.

Chapter **10**

Metal Roofing

Galvanized, ribbed sheet-metal panels and aluminum shakes and shingles are among the most durable roofing materials. Although such materials are more expensive than asphalt fiberglass-based shingles, metal roofing will last far longer than shingles. Comparisons depend on the weight of the fiberglass shingles and the anticorrosion properties of the metal roofing installed.

Although some sheet-metal materials require periodic painting, keep in mind that prepainted galvanized steel roofing is now available in a wide variety of patterns and colors. With proper maintenance, metal roofing will not have to be replaced for 50 to 75 years. During that time, asphalt fiberglass-based shingles will need to be replaced two to three times.

SHAKES AND SHINGLES

Aluminum shakes and shingles vary in style and installation requirements; detailed application instructions are provided by manufacturers. The advantages of using aluminum roofing include long life, low maintenance, light weight, and attractive appearance. Because aluminum reflects up to 80 percent of the radiant heat from sunlight, your summer house-cooling requirements will be reduced. In the winter, snow and ice are quickly shed as the warmed aluminum will encourage the snow and ice to slide off the roof surface.

The aluminum shakes manufactured by Reinke Shakes, Inc. must be installed similar to wood shakes (see chapter 9). An 18-inch-wide interlay strip of 15-pound felt must be applied beneath each course of shakes. As with wood shakes, applying these interlay strips adds to the time required for installation of the shakes.

Reinke shakes can be applied over new plywood, particleboard, chipboard, or similar sheathing material. The shakes can be applied over one layer of asphalt shingles or over one layer of wood shingles (but not wood shakes), but you will first have to install suitable sheathing over the worn roofing material so that the shakes will have a solid nailing base.

See chapter 3 for guidelines for estimating the number of square feet of roof coverage. To begin installing the shakes, nail down a $2^1/_2$- × -$^1/_8$-inch furring strip from rake to rake and flush with the eaves. Because the factory-drilled nail holes are located on the exposed portion of the shakes, three layers of 15-pound felt must be installed with the shakes.

At the eaves, install two layers of 18-inch-wide felt. The second layer of felt should be lapped 2 inches higher than the first layer. (See FIG. 10-1.)

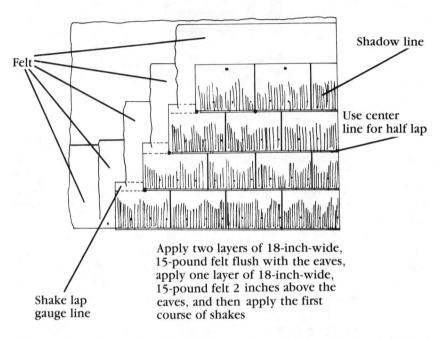

Shadow line

Felt

Use center
line for half lap

Apply two layers of 18-inch-wide,
15-pound felt flush with the eaves,
apply one layer of 18-inch-wide,
15-pound felt 2 inches above the
eaves, and then apply the first
course of shakes

Shake lap
gauge line

10-1   Install two layers of 18-inch-wide felt over the roof deck, and an interlay strip of felt with each course of shakes.

The interlay felt strip beneath each course of shakes makes up the third felt layer. The three felt layers will seal around the nails. Moisture that manages to penetrate the nail holes will flow over the top of the next course of shakes.

Working left to right, apply the starter row of shakes. To match the corrugations on the starter course, and therefore, subsequent courses of shakes, cut off the shadow line of one shake, turn the shake upside down, and half-lap the shake along the first shake to be installed.

Install the interlay felt strip. Stagger the alignment of the first shake of the second course by half the width of a shake so that every other course of shakes will be aligned.

There are two options for installing Reinke shakes along the rakes of a roof. One method is to install drip edge at the rakes (chapter 3) and to apply the shakes either flush with the drip edge or overhanging the edge by 1 inch. The second method is to not use drip edge, but to position and then trim each shake so that it extends 1 inch over the rake. Bend down the rake edge of the shake about 1 inch and then nail the shake in place (FIG. 10-2). The edge is bent in order to make the rake watertight. Drip edge does the job better; it will give a straighter, more attractive appearance, and it requires less work.

To install shakes at a valley (FIG. 10-3), first apply two layers of 15-pound, 36-inch-wide felt down the length of the valley. Install a length of aluminum valley material by driving nails only at the edges of the valley. See chapter 7.

As you install the shake courses at the valley, apply silicone caulking under each shake. Using tin snips, trim the shakes along the corrugations. Use a power saw to cut across the shake corrugations.

To apply ridge or hip capping, first install a folded 18-inch strip of felt (9 inches total width) along the hip or ridge. Use a utility knife to cut off the shake shadow line to make the caps. Use the standard lap and coverage for the capping (FIG. 10-4).

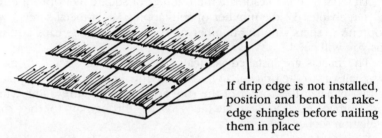

If drip edge is not installed, position and bend the rake-edge shingles before nailing them in place

10-2 Position, bend, and then install the shakes at the rakes. An alternative method is to use drip edge at the rakes.

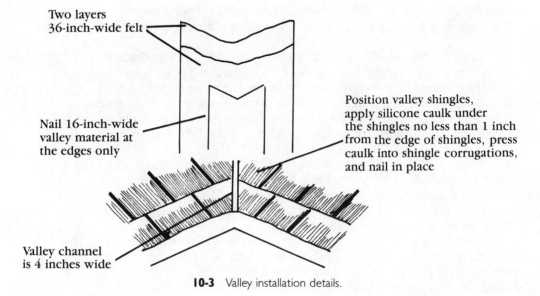

Two layers 36-inch-wide felt

Nail 16-inch-wide valley material at the edges only

Position valley shingles, apply silicone caulk under the shingles no less than 1 inch from the edge of shingles, press caulk into shingle corrugations, and nail in place

Valley channel is 4 inches wide

10-3 Valley installation details.

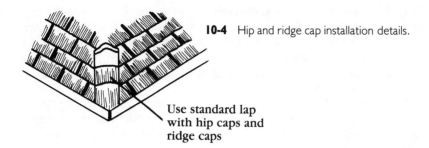

10-4 Hip and ridge cap installation details.

Use standard lap
with hip caps and
ridge caps

SHEET METAL

Most galvanized sheet-metal roofing, such as Channeldrain by Wheeling-Pittsburgh Steel, Wheeling Corrugating Co., has attractive ribbed seams. Terne-coated stainless steel, by Follansbee Steel Corporation, is a similar product. The steel sheets are coated with molten zinc by a process known as *hot-dip galvanizing*. The zinc provides corrosion resistance. For residential structures, the metal thickness should be about 26 gauge.

The first step is to estimate the number of square feet of roof coverage (see chapter 3), the number of self-tapping sheet-metal screws with neoprene washers, and the number of 10-foot-long ridge caps and rake caps you will need.

The panels are installed vertically in sheets that stretch from the eaves halfway to the ridge of a typical residential home. Begin by placing the first panel at the corner of the roof away from the prevailing winds. Work from the eaves to the ridge. (See FIGS. 10-5 through 10-9 for the proper installation sequence.)

Channeldrain's 3-foot-wide, 13-foot-long panels can be installed quickly once the first panel has been carefully and properly aligned and positioned at the eaves with the aid of precisely marked chalk lines. Be extremely careful that the first panel is square. The proper alignment of the remaining sheets depends upon the correct placement of the first panel.

Once the first panel is aligned, you can use a drill to install screws every 2 to 4 feet. Channeldrain panels are secured at the ridge with caps and at the rakes with end caps. At the eaves, the sheets are pulled together at the ribs and secured with short screws. Always set fasteners flush with the panel surface. Never overdrive nails or screws and do not dimple the steel. Subsequent sheets are installed quickly until you reach an obstacle such as a vent pipe, a chimney, or a valley.

To cut or trim panels, it is essential that you use an electric saw with a steel cutting blade or a Carborundum blade. Place the panel exterior side down so that the exterior surface will not be marred. Carefully brush off any metal particles and filings that otherwise could cause rust marks or bleeding on the installed panel surfaces.

For a vent pipe, cut a circular hole about $1/2$ inch larger than the pipe, and make a horizontal slot long enough for the flange to fit. The "top" of the flange must slide under the sheet-metal panel. The "bottom" of the flange fits over the pipe and is installed on top of the panel (FIG. 10-10).

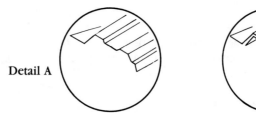

Detail A   Detail B

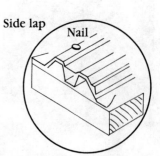

Side lap   Nail

10-5 Detail A shows how panel I is aligned with the rake and eaves. Detail B shows how panel I and panel 3 are lapped. Side lapped panels must be nailed so that the panels are securely drawn together. Do not overdrive the fasteners or dimple the steel.

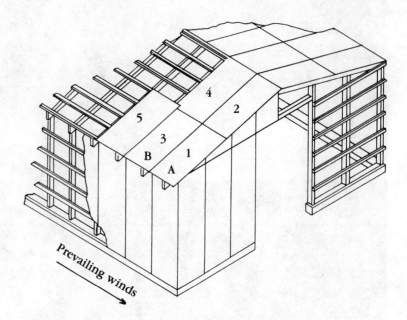

Prevailing winds

With middle-of-the-roof chimneys, you will have to position and then cut at least two sheet-metal panels to fit around first one side and then the other side of the chimney. Rake chimneys are much easier to handle. Basically, all you will have to do is measure and cut the area to be removed from a panel, install the sheet, and flash the chimney area. See the section on chimneys in chapter 7.

Installing panels at valleys requires careful cutting and trimming to obtain a watertight, as well as a neat, appearance. Install the valley material as described in chapter 7, but note that no border shingles are required with sheet-metal roofing. The sheet-metal panels must extend over the valley surface. Do not drive screws closer than 3 inches from the edge of the valley material.

10-6 Install the first panel with a 1-inch overlap at the eaves and the rake.

10-7 The proper alignment of the first panel will allow you to correctly install the additional panels.

10-8 Position the remaining panels to allow for the correct amount of overlap.

10-9 Use tin snips to cut counterflashing. This flashing must be installed flush with the wall and on top of the sheet metal panels.

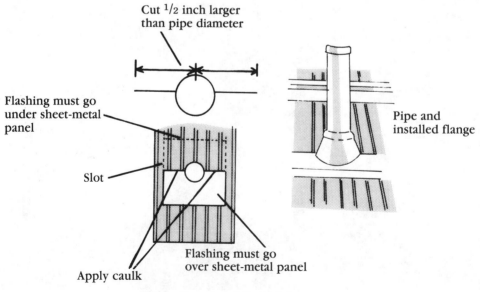

Cut ¹/₂ inch larger
than pipe diameter

Flashing must go
under sheet-metal
panel

Slot

Apply caulk

Flashing must go
over sheet-metal
panel

Pipe and
installed flange

10-10 Flange installation details.

Chapter **11**

Corrugated Asphalt Roofing

Because asphalt impregnated, corrugated roofing sheets and tiles are lightweight, strong, and easy to install—the corrugations help make them almost self-aligning—Onduline roofing material (FIG. 11-1) can be a good choice for do-it-yourselfers. Available in various colors and in either a granulated or a smooth finish that can be repainted by brush or by spraying with 100 percent acrylic latex exterior paint, the sheets and tiles have been assigned a class C fire rating by Underwriters Laboratories.

Onduline roofing sheets or tiles can be applied over a layer of felt on a new plywood roof deck or over a layer of worn asphalt shingles. If you are reroofing over metal roofing, Onduline recommends using sheet roofing rather than tiles. When applied over a layer of worn asphalt shingles, Onduline roofing needs no special preparations. If you have one layer of metal roofing with standing seams, you must flatten the seams before you install the sheets. With corrugated metal roofing, you first will have to apply wood cross strips, called *purlins,* no wider than 24 inches on center, as a nailing base. (See FIG. 11-2.)

Onduline sheets and tiles can be installed on a 3-in-12-foot-or-steeper slope. Sheets always must be installed over purlin support strips that are typically spaced 24 inches on center. Tiles can be applied to a similar slope over a solid deck of plywood or over one layer of worn roofing. Special installation requirements, such as placing the purlins 18 inches on center, should be followed if your location has unusually heavy snow accumulation. If your roof deck is more than 80 feet from eaves to the ridge line, sheets should not be used.

While the manufacturer suggests that Onduline roofing can be applied over two worn layers of asphalt shingles, keep in mind that most residential structures are not designed to support three layers of roofing materials. In addition, the nails used to install the new roofing must adequately penetrate the roof deck through more than two layers of shingles.

11-1 Onduline roofing products can be installed over one layer of worn roofing or on a new deck.

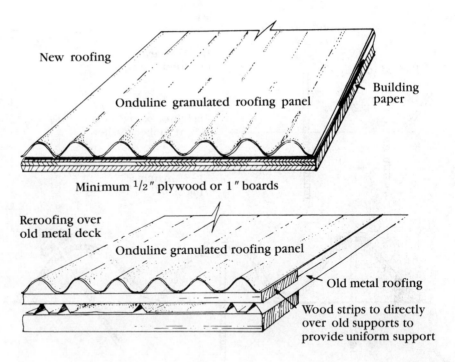

New roofing

Onduline granulated roofing panel

Building paper

Minimum 1/2" plywood or 1" boards

Reroofing over old metal deck

Onduline granulated roofing panel

Old metal roofing

Wood strips to directly over old supports to provide uniform support

11-2 Onduline panels can be applied over felt on a new deck, where worn roofing has been removed, or over a layer of worn roofing. To reroof over a layer of metal roofing, purlins must first be installed as a nailing base.

According to Onduline, nails should penetrate 1 inch through the deck. (See FIG. 11-3.)

The installation of Onduline roofing will create a dead air space between the new and old roofing material. This dead air space will provide a barrier to the transfer of heat during the summer and provide additional insulation during the winter.

Tools you will need are a claw hammer or a roofing hatchet, a chalk line, a nail apron, a tape measure, a utility knife and blades, and an electric saw with a carbide-tipped blade. A utility knife can be used to cut sheets or tiles parallel with the corrugations. Use a saw to cut across corrugations. Remember to wear safety eyeglasses while cutting across corrugations. A powered nail gun can be used to install tiles, but do not attempt to use a staple gun on either Onduline sheets or tiles.

Step-by-step illustrations and directions for applying sheets and tiles on rectangular sections, hips, valleys, out-of-square sections, and how to install flashing (FIG. 11-4) and skylights (FIG. 11-5) are detailed in the installation brochures that come with the materials. Check with your local building suppliers and Onduline, at the address provided in the Sources appendix, for more information on the availability of corrugated asphalt roofing products.

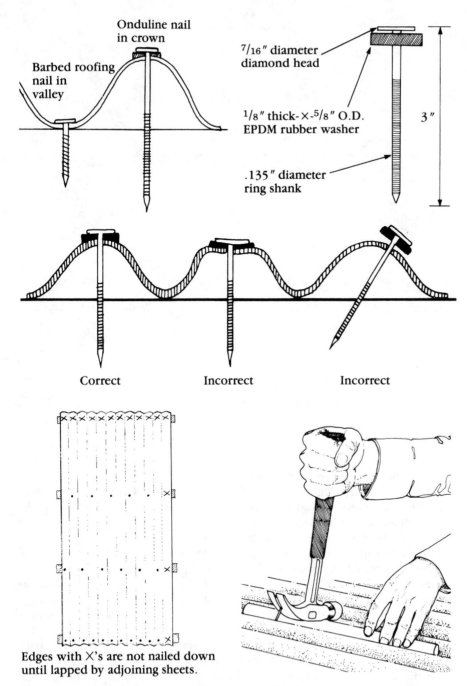

Onduline nail
in crown

Barbed roofing
nail in
valley

$7/16''$ diameter
diamond head

$1/8''$ thick-×-$5/8''$ O.D.
EPDM rubber washer

$3''$

.135" diameter
ring shank

Correct   Incorrect   Incorrect

Edges with X's are not nailed down
until lapped by adjoining sheets.

11-3 Barbed roofing nails are driven through tile valleys $4^1/2$ inches from the top edge of
the tile so that the courses lap the nail heads. Drive 3-inch Onduline nails through corrugation crowns at the eaves, at the ridge cap crowns, and at the lower corners and lower
center of the tiles. If you use 4-inch nails, up to $1^1/2$ inches of rigid insulation can be installed
beneath Onduline roofing.

ESTIMATING MATERIALS

Use the information outlined in chapter 3 and the Onduline guidebook to help you determine your roof's total number of square feet. Remember to include 10 percent for waste. Again, consult with your local Onduline dealer for details.

Sheets Each sheet of Onduline roofing covers 26 square feet. Because the sheets must be overlapped when installed, it takes $4^1/2$ sheets to cover 100 square feet of roof surface. You will need 24 nails to install each sheet and 38 nails for each ridge cap. To find the total number of closure strips you will need, first determine the total length—in inches—of all the eaves and twice the length of the hips, valleys, and rakes. Divide this total by 44 (the length in inches of one closure strip) to obtain the number of strips needed.

Tiles Each tile of Onduline roofing covers $6^1/2$ square feet. Because tiles must be overlapped when installed, 24 tiles will cover 100 square feet of roof surface. To install each tile, you will need 12 barbed roofing nails. In addition, you will need 12 galvanized, ring-shanked, washered nails for each tile. Each ridge cap provides 6 feet of coverage. The total number of closure strips needed can be determined by finding the length, in inches, of all the eaves and twice the length of hips, valleys, and rakes. Divide the total by 44 (the length of one strip) to find the number of strips needed.

SHEET APPLICATION

To prepare for the installation of Onduline sheets, follow the guidelines described in chapters 3 and 4. Prior to nailing on the first sheet, install drip edge along all of the rakes and eaves. The next step is to establish a uniform overhang at the rakes and at the eaves. The sheets should be installed flush with the drip edge at the rakes, and the sheets should overhang the eaves by $1^3/4$ inches.

If you are right-handed, start at the lower left corner—of the back of your roof—in order to keep the work in front of you. If you are left-handed, begin work at the lower right corner of the back of the roof. The following descriptions are oriented for right-handed persons in order to simplify the instructions. Reverse the starting points—to the opposite side of the roof sections—if you are left-handed.

Starter Course At the left-hand corner of the rake and eaves, measure in 48 inches—remember to align the sheets with the drip edge—and make a mark. At the ridge, measure in the same length and make another mark. Snap a chalk line between these marks to provide proper sheet alignment. If you are working alone, you can loop the hook at the end of the chalk line over a nail head and then snap the chalk line. (See FIG. 4-35.)

To provide a $1^3/4$-inch overhang at the eaves, at the left-hand corner of the rake and eaves, measure $77^1/4$ inches up from the eaves. Measure the same distance at the right-hand corner and snap a chalk line between the marks.

Position the first sheet of roofing at the left-hand corner of the rake and eaves. At the purlins along the rake, drive washered nails only

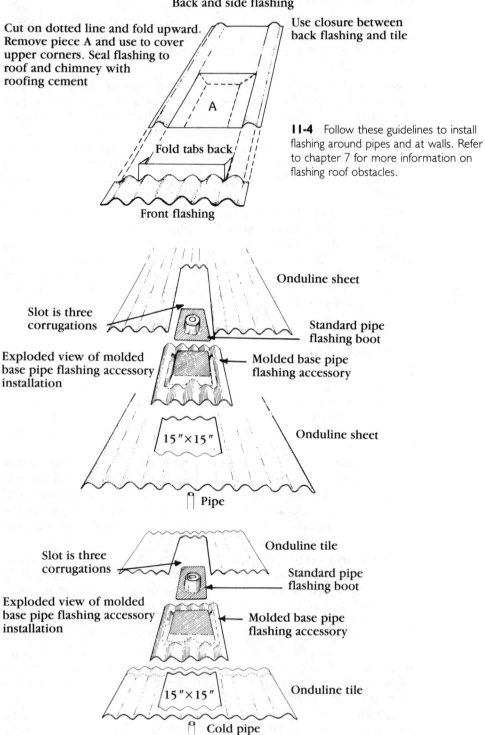

Back and side flashing

Cut on dotted line and fold upward.
Remove piece A and use to cover
upper corners. Seal flashing to
roof and chimney with
roofing cement

Use closure between
back flashing and tile

A

Fold tabs back

11-4 Follow these guidelines to install
flashing around pipes and at walls. Refer
to chapter 7 for more information on
flashing roof obstacles.

Front flashing

Onduline sheet

Slot is three
corrugations

Standard pipe
flashing boot

Exploded view of molded
base pipe flashing accessory
installation

Molded base pipe
flashing accessory

Onduline sheet

15″×15″

Pipe

Onduline tile

Slot is three
corrugations

Standard pipe
flashing boot

Exploded view of molded
base pipe flashing accessory
installation

Molded base pipe
flashing accessory

15″×15″

Onduline tile

Cold pipe

11-4 Continued.

Use butyl caulking at all flashing joints and closure strips

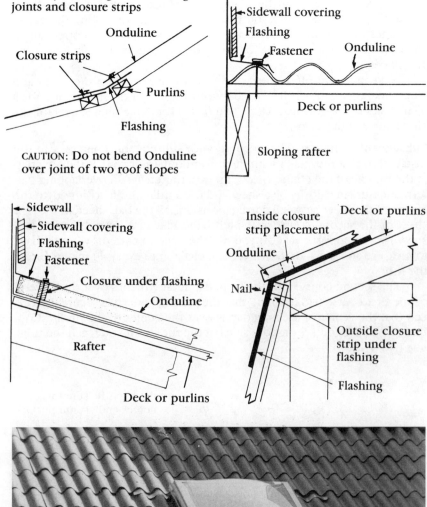

CAUTION: Do not bend Onduline over joint of two roof slopes

11-5 Where an Onduline skylight is installed, use butyl caulk to seal the overlapping skylight panels and roofing sheets.

through the top of corrugations. Do not install nails along the top or right-hand side of the first sheet until the next sheet is positioned, properly overlapped, and ready to be fastened. If you are using closure strips, install the first strip before you fasten the corrugations along the eaves. Now check for the proper installation of the first sheet. The correct alignment of the remaining sheets depends upon accurate installation of the first sheet.

Next, measure off a series of 44-inch marks at both the eaves and ridge line. Snapping chalk lines between these marks will provide for a 1-inch sidelap of corrugations as you install the remaining sheets for the first course at the eaves.

Subsequent Courses The second and subsequent courses can be installed so that they align with the lower courses or so that they overlap at the middle of the course below. In order to create an overlapping pattern, measure in halfway on a sheet and use a utility knife to cut down the length of the sheet between corrugations. Install the half sheet, as the first sheet of the second course at the left-hand rake, and save the other half sheet for the opposite rake. If you use the overlap pattern, you will have to measure and snap additional vertical chalk lines every 44 inches across the roof.

The second course must be lapped at least 7 inches over the top end of the first course (FIG. 11-6). So that the nailing pattern is secure, be certain that the center of each overlap is over the center of the purlin. Continue installing courses on both sides of the roof until you reach the ridge line (FIG. 11-7).

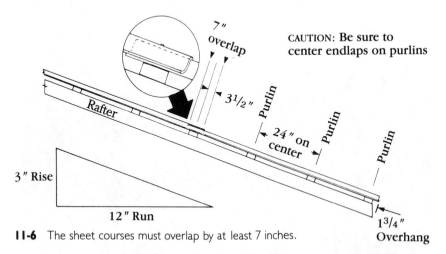

11-6 The sheet courses must overlap by at least 7 inches.

At the ridge, the top ends of the sheets from both sides of the roof should be within 2 inches of each other, in order to support the ridge capping. Instead of installing ridge capping, consider adding a continuous ridge ventilator. If you do use ridge capping, position the first cap away from prevailing winds and 3 to 6 inches from the ridge end. Install a closure strip and drive washered nails through the cap, the closure, and the

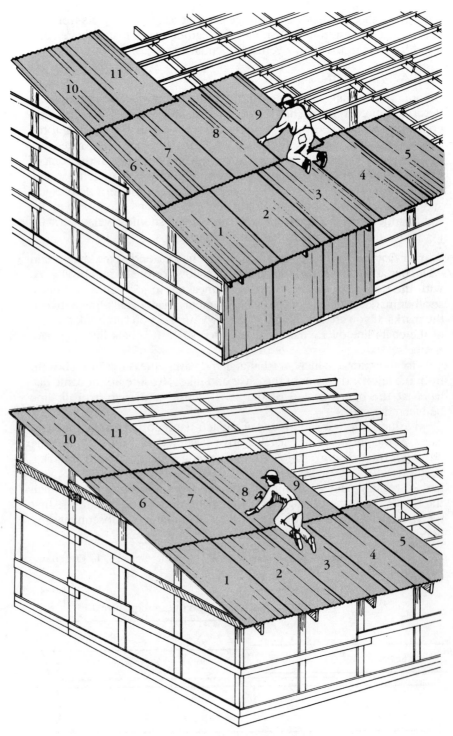

11-7 The sheet installation pattern.

underlying sheets. Next, cut the ridge cap portion that projects from the rake so that the resulting fold provides a weather guard. Install the remaining ridge caps so that they overlap by 7 inches.

TILE APPLICATION

Before you install Onduline tiles, read the preparation guidelines described in chapters 3 and 4. Prior to installing the first tile, nail drip edge along all of the eaves and rakes. Allow for a 1-inch overhang at the eaves and plan to align the tiles flush with the drip edge at the rakes.

If you are right-handed, begin at the lower left corner of the back of the roof in order to keep the work in front of you. If you are left-handed, start working at the lower right corner of the back of the roof. In order to simplify the instructions, the following descriptions are oriented for right-handed persons. If you are left-handed, you should reverse the starting points to the opposite side of the roof sections.

Starter Course   At the left-hand corner of the rake and eaves, measure in 48 inches and make a mark. Remember to allow for the tiles to be aligned with the drip edge. At the ridge, again measure in 48 inches and make another mark. To ensure proper tile alignment, snap a chalk line between the marks. If you are installing the tiles by yourself, you can hook the end of the chalk line over a nail head and then snap the chalk line. (See FIG. 4-35.)

To allow for a 1-inch overhang at the eaves, measure 78 inches up from the left-hand corner of the eaves and rake. Measure up the same distance at the opposite end of the roof section and snap a chalk line between the marks.

Position the first roofing tile at the left-hand corner of the rake and eaves. At the purlins along the rake, drive washered nails only through the crown of the first corrugation so that the edge is aligned with the rake. Position a closure strip between the drip edge and the tile (FIG. 11-8). Using barbed roofing nails positioned $4^{1}/_{2}$ inches from the top of the tile, nail the tile between the corrugation. Now is the time to make sure the first tile has been installed properly. Correct alignment of subsequent tiles depends upon proper installation. (See FIG. 11-9.)

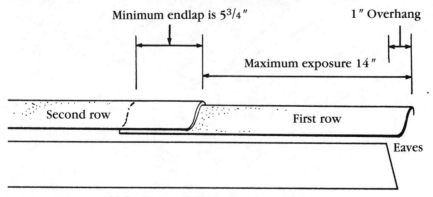

Minimum endlap is $5^{3}/_{4}$"

1" Overhang

Maximum exposure 14"

Second row   First row   Eaves

11-8   Tile installation at the rake and eaves.

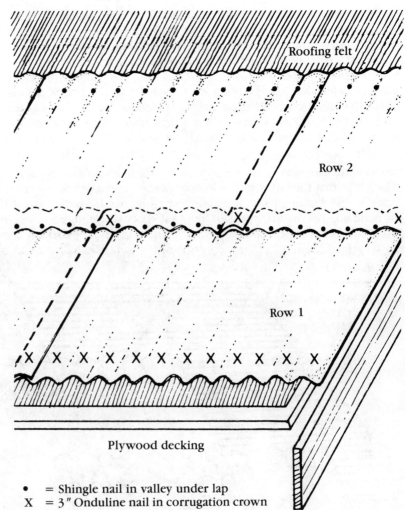

11-9 Alignment of the tiles at the eaves and rake, and the nailing pattern for the first two courses.

Roofing felt

Row 2

Row 1

Plywood decking

• = Shingle nail in valley under lap

X = 3″ Onduline nail in corrugation crown

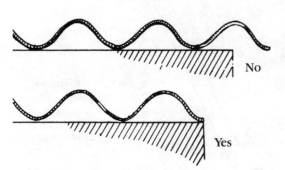

11-10 The corrugations must overlap by at least 1 inch.

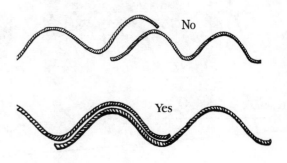

11-11 Adjust the sidelap for proper alignment at the rake.

Measure off a series of 44-inch marks at both the ridge line and at the eaves. Snap chalk lines between these marks to ensure a 1-inch sidelap (FIG. 11-10) of corrugations as you install the subsequent tiles. Allowing for a 1-inch sidelap, install all but five of the remaining tiles of the first course. Position, but do not nail down, the remaining five tiles at the eaves in order to determine whether or not the last tile in the course will align with a "valley" between tile corrugations. Adjust the amount of sidelap for the last five tiles by changing the sidelaps until the proper alignment at the rake is achieved (FIG. 11-11). When the alignment is correct, nail the last five tiles in place.

Subsequent Courses   The second and subsequent courses must be installed so that they overlap the middle of the course below. In order to create an overlapping pattern, measure in 14 inches from the rake at the left-hand edge of the first course. Count off six corrugated crowns on the first tile of the second course (FIG. 11-12). Use a utility knife to cut down the length of the tile between corrugations. Install the half tile—as the

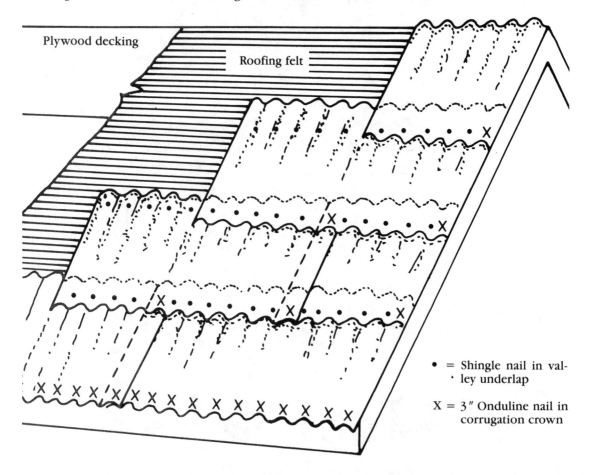

Plywood decking

Roofing felt

• = Shingle nail in val-
ley underlap

X = 3″ Onduline nail in
corrugation crown

11-12   Note the nailing patterns and the overlapping of the courses.

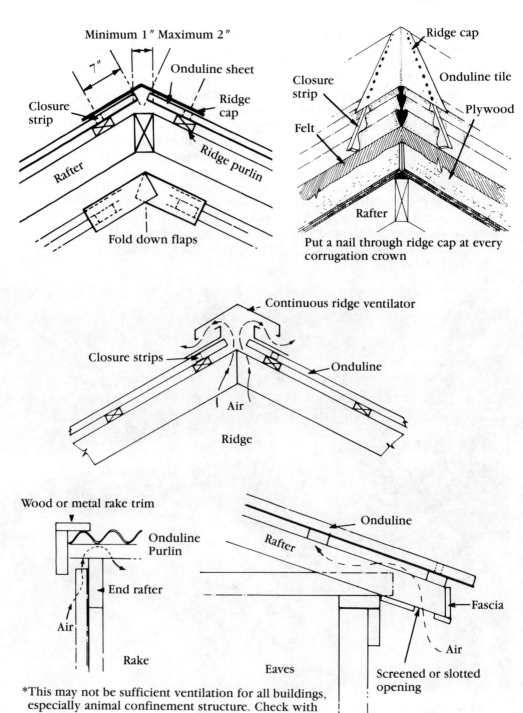

Minimum 1″ Maximum 2″

7″

Closure strip

Onduline sheet

Ridge cap

Rafter

Ridge purlin

Fold down flaps

Ridge cap

Closure strip

Onduline tile

Felt

Plywood

Rafter

Put a nail through ridge cap at every corrugation crown

Continuous ridge ventilator

Closure strips

Onduline

Air

Ridge

Wood or metal rake trim

Onduline

Purlin

End rafter

Air

Rake

Onduline

Rafter

Fascia

Air

Eaves

Screened or slotted opening

*This may not be sufficient ventilation for all buildings, especially animal confinement structure. Check with an engineer.

11-13 Ridge line ventilation details.

first tile of the second course at the left-hand rake—and save the other half of the tile so that it can be installed at the opposite rake. Using the first half tile as a starting point, measure and snap additional vertical chalk lines every 44 inches across the roof section.

The second course must be lapped at least 14 inches over the top end of the first course. Continue installing courses on both sides of the roof until you reach the ridge line.

At the ridge, the top ends of the sheets, from both sides of the roof, must be within 2 inches of each other. Consider adding a continuous ridge ventilator instead of ridge caps (FIG. 11-13). If you do use ridge capping, position the first cap away from prevailing winds and 3 to 6 inches from the ridge end. Install a closure strip and drive washered nails through the cap, the closure, and the underlying tiles. To provide a weather guard at the rake, cut and fold the ridge cap portion that projects from the rake (FIG. 11-14). Install the remaining ridge caps so that they overlap by 7 inches.

11-14   Closure strips and ridge cap installation provide wind and weather protection at the roof edges.

HIPS AND VALLEYS

For hip roofs, courses of Onduline sheets or tiles must be trimmed along each side of the hip so that the courses provide support for ridge capping installed over the hip. To determine if you have the proper coverage at the hip, measure the width of the capping, divide by two, and measure the result from each side of the center of the hip. Use these marks to snap chalk lines on each side of the hip. Before installing the hip capping, make certain that you have adequate framing support as a nailing base along the hip and at the ridge line. Cut two-corrugation-wide closure strips, and use butyl caulking to install a two-corrugation-wide overlapping pattern. (See FIG. 11-15.) Install a 7-inch end lap at the bottom of the hip.

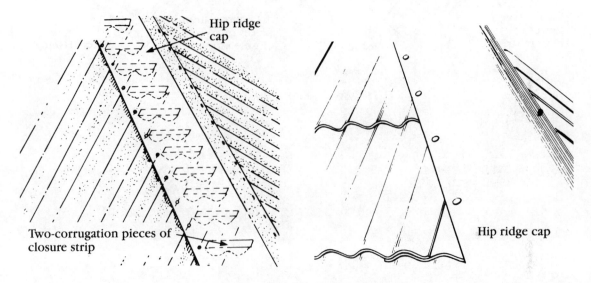

Hip ridge cap

Two-corrugation pieces of closure strip

Hip ridge cap

11-15 Hip capping installation details.

For valley flashing, nail in place a minimum of 36-inch-wide sheets. If your valley flashing is not long enough to cover the full length of the valley, allow 6 inches for each overlap between sections. Refer to chapter 7 for additional information on valleys and other roof obstacles.

Position and trim, but do not nail in place, the Onduline sheets or tiles at an angle to allow for a 6-inch-wide center channel. Turn over the cut sheets or tiles and use butyl caulk to attach two-corrugation-wide pieces of closure strip in an overlapping pattern. See FIG. 11-16. Add a 1/2-inch bead of caulking on the flat side of the closure strips and on the sheet between the closure strips. Turning the sheets or tiles upright, position the roofing so that the caulked closures meet the valley flashing. Now nail the sheets or tiles in place (FIG. 11-16).

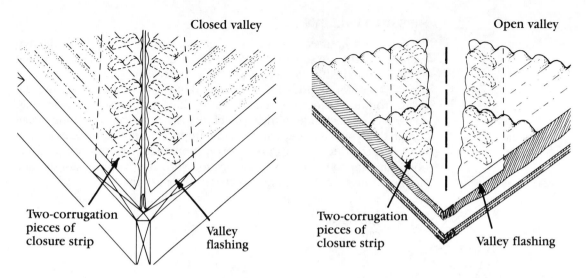

11-16 Valley flashing installation details.

Chapter 12

Ventilation

*P*roper attic ventilation can lower the cost of cooling your home by 60 percent, prolong the life of roofing materials, and prevent the premature deterioration of attic-insulation materials. Some 70 percent of American homes are not designed and constructed with proper ventilation in mind. Fortunately, installing wind-driven turbines, powered ventilators, roof-line louvers, cupolas, ridge vents, skylights, and whole-house fans are easy do-it-yourself projects. The ideal time to install ventilating units is during the shingling of a new roof.

A BALANCED SYSTEM

When hot air rises to your home's attic, it will remain trapped there unless you have a good attic-ventilation system (FIG. 12-1). Without adequate ventilation, summer weather can easily superheat attic air to 140 degrees. Through convection, the heated attic air will raise the temperature of your living areas. The trapped, overheated air will radiate warmth downward along the ceiling, walls, and joists. The very hot air keeps warm air in your living areas from rising and adds to the burden on your air-conditioning system. To overcome this heat, your air conditioner will have to operate almost constantly during summer weather (FIG. 12-2). Superheated attic air will actually scorch rafter boards and sheathing, wilt insulation, and cause shingles to cup and buckle.

Many newer homes—designed with insulation, weatherstripping, doors, and windows that are designed to reduce energy losses—are more susceptible to condensation problems than older homes that have not been retrofitted. During winter, an attic must have more ventilation than during summer because windows and doors are usually kept closed in cold weather. Unless water vapor—produced by the use of bathtubs, showers, and home appliances—that passes through the ceiling and collects in the attic is removed by adequate ventilation, it will soak and destroy insulation and perhaps even rot rafters.

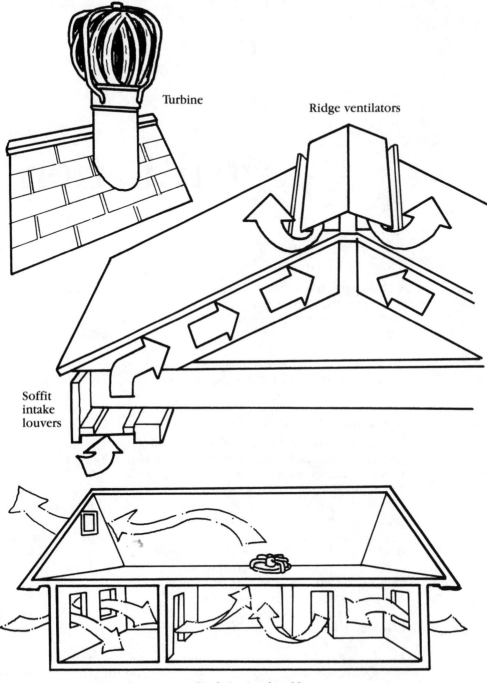

Turbine

Ridge ventilators

Soffit
intake
louvers

Cool air circulated by
a whole-house fan

12-1   A balanced ventilation system.

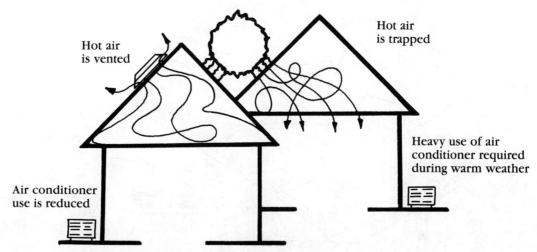

Hot air
is vented

Hot air
is trapped

Heavy use of air
conditioner required
during warm weather

Air conditioner
use is reduced

12-2 Without adequate attic insulation, your air-conditioner will have to work continuously during summer weather.

Ice Dam Formation

In some areas of the country, ice dams form on the roof edges of poorly ventilated homes (FIG. 12-3). An ice dam will form when the heat from trapped attic air slowly melts snow on the middle and upper portions of a roof section. During very cold but sunny days, radiant heat will increase the flow of meltwater. As the water travels toward the edges of the roof, it cools because there is less heat or no heat from the attic and there is more snow at the eaves. Some of the water will freeze and create a dam of ice.

As the meltwater increases, so does the accumulation of ice that eventually will force additional meltwater under the shingles and through the sheathing and into an attic or a wall. An ice dam can create enough force to lift some shingles, but most of the damage will not be found on the roof surface. The water will penetrate insulation materials and the ceiling. By the time you notice spots on the ceiling and peeling paint or plaster cracks, significant damage inside the home will have occurred.

Proper ventilation, along with at least an adequate amount of insulation, will prevent the uneven temperatures that cause ice dams. Because warm air rising through the attic or ceiling of the house is a key factor in ice dam formation, you should reduce the heat flow by adding insulation and by increasing the number of attic ventilators. More cold air in the attic will decrease the amount of meltwater because the bottom of the roof will approach the outside temperature. The additional insulation will reduce the flow of heat rising from the living area.

Alternative solutions to ice dam problems include installing a continuous strip of sheet metal on top of the shingles at the eaves. On steeply sloped roofs, the slick metal surface and gravity will combine to discourage the formation of ice. Also, you could install heat cables that are designed to be clipped at a zigzag angle to the eaves' shingles. The cables can be extended into the gutters to ward off freeze-ups.

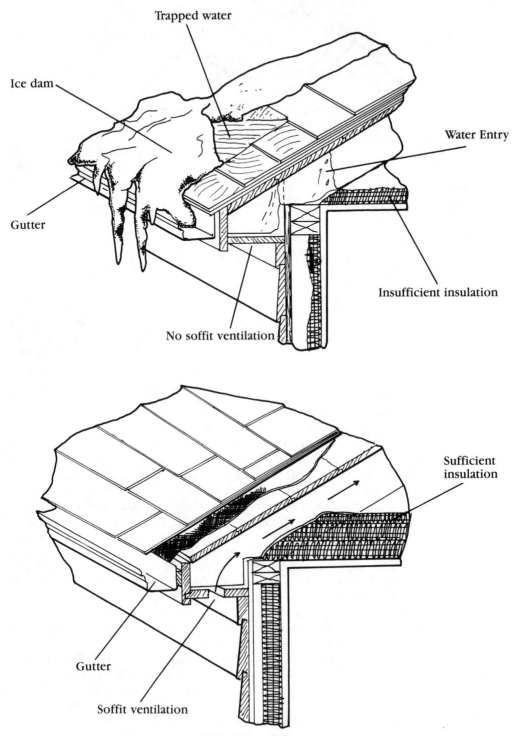

12-3 Ice dam formation.

On a new roof deck or where the old shingles have been torn from the surface, additional protection from ice and meltwater at the eaves can be obtained by installing a 3-foot-wide strip of roll roofing, selvage edge, or—best of all—a water and ice shield membrane that will seal around the nails used to secure the shingles installed along the eaves. Several manufacturers offer rubberized asphalt and polyethylene sheets that are designed to bond directly to the roof deck and provide a backup barrier against meltwater from ice dams. Check with your local building supplier to see if the application of water and ice shield membrane is recommended for your area. Also, keep in mind that this type of membrane makes an excellent underlayment for valleys.

Types of Ventilators

Roof and gable ventilators are designed to let superheated attic air escape and prevent rain, snow, and insects from entering. Whole-house fans can be mounted in attic floors or in walls.

Roof-Line Louvers Roof louvers (FIG. 12-4) are simple devices that cover holes cut in the roof near the peak. Hot air escapes as it rises.

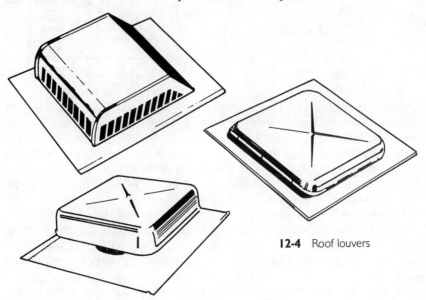

12-4 Roof louvers

Gable-End Vents Gable vents (FIG. 12-5) are probably the most common type of ventilation unit. They are usually triangular or square and are flush or recess mounted.

Under-Eaves Vents Under-eaves soffit ventilators (FIG. 12-6) are essential for a balanced ventilation system. About 2 square feet of soffit ventilation is needed for every 600 square feet of attic-floor area.

Wind-Driven Turbines A turbine vent (FIG. 12-7) is designed to be driven by the wind from any direction. As the turbine spins, reduced air pressure in the throat of the unit draws hot, humid air from the attic.

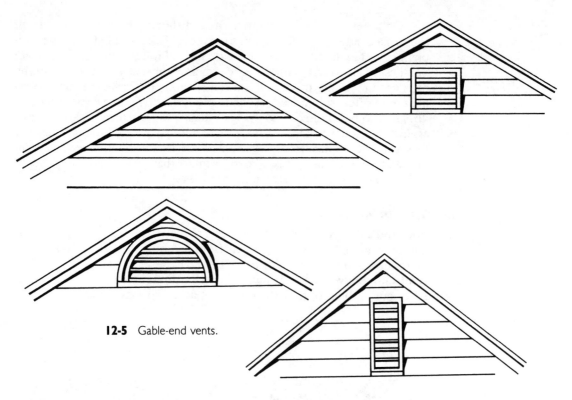

12-5 Gable-end vents.

Cupolas Cupolas (FIG. 12-8) are essentially disguised large, wind-driven ridge ventilators. With a weather vane, a cupola can make an attractive roof structure.

Ridge Ventilators A ridge vent (FIG. 12-9) uses the natural flow from the eaves to the top of the ridge to exhaust superheated air from the attic. When combined with under-eaves vents and a whole-house fan, ridge vents provide a completely balanced ventilation system.

Whole-House Fan A whole-house fan (FIG. 12-10) is designed to help cool a house by drawing cooler air throughout the house during early morning hours and evening hours. A whole-house fan system combined with proper attic ventilation can reduce air-conditioning costs by 60 percent or more.

A cooler roof will last longer than one where the shingles are continually subjected to the expansion and contraction caused by the constant superheating of attic air throughout summer months. To vent superheated attic air (that can reach 50 degrees higher than the outside temperature), install turbines—one 12-inch unit for about every 600 square feet of attic floor—near the peak of the roof or ridge vents mounted along the full length of the peak. Properly installed turbines or ridge vents will keep an attic within 5 to 10 degrees of the outside air temperature.

In temperate climates, turbines or ridge vents in combination with soffit vents can be used with a whole-house fan to virtually eliminate the

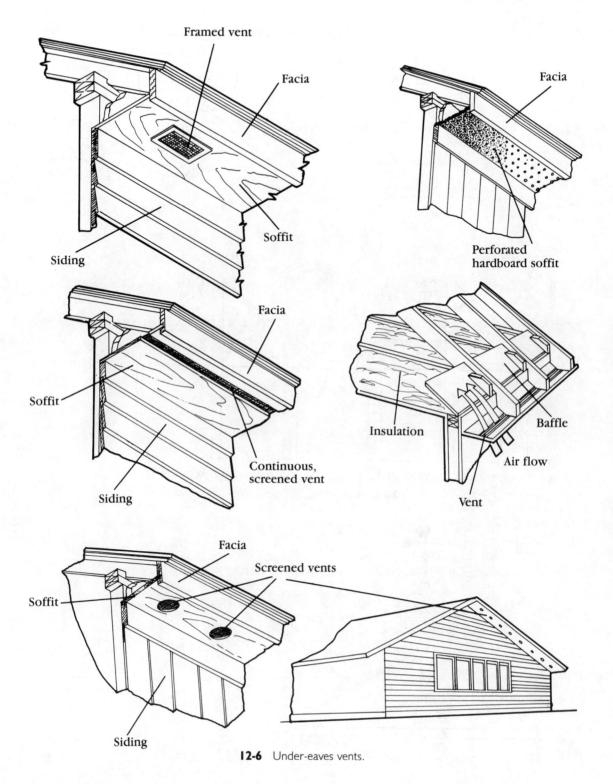

12-6 Under-eaves vents.

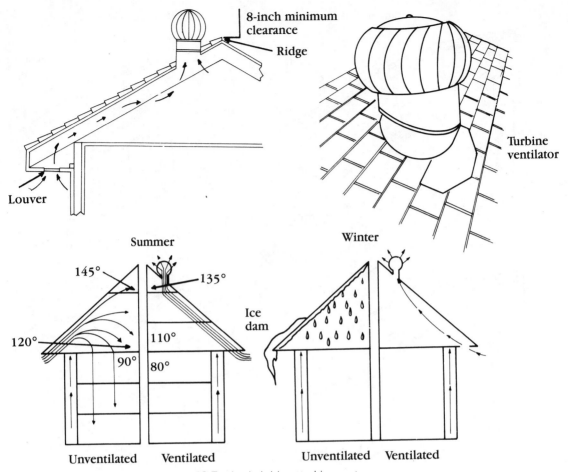

12-7 A wind-driven turbine vent.

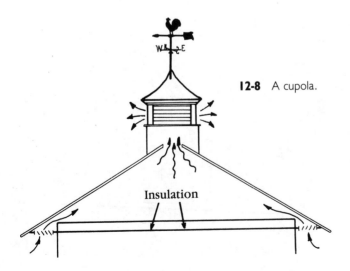

12-8 A cupola.

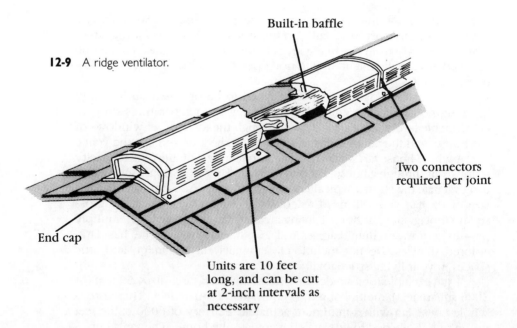

12-9 A ridge ventilator.

Built-in baffle

Two connectors
required per joint

End cap

Units are 10 feet
long, and can be cut
at 2-inch intervals as
necessary

12-10 A whole-house fan.

need for air-conditioning, except during the hottest summer weather. The whole-house fan must be installed in the attic floor in order for the fan to draw air directly from the living areas below. A whole-house fan can be installed in a wall, but it might not operate as efficiently as a floor-mounted unit.

A whole-house fan system is designed to be operated during early morning hours and evening hours when outside temperatures drop to comfortable levels. Cool air is drawn through the lower-level windows of the house, and the fan pushes the air through the attic and out the vents.

A whole-house fan is rated by its cubic-feet-per-minute (cfm) capacity to move air through a house. Properly installed, a whole-house fan system can evacuate the air in a typical house in 2 minutes. You can calculate the size of the fan you will need to cool your home by totaling the gross square footage of your home's living areas and multiplying by 3; multiply by 4 in very warm, humid areas. Add 15 percent if your home has dark-colored shingles. Do not include closet, stairwell, basement, and attic spaces in your living-space total.

As an example for a single-story home that measures 40 × 50 feet (or 2000 square feet), multiplying by 3 equals 6000 square feet. Therefore, a whole-house fan with a minimum actual air delivery of 6000 cubic feet per minute is needed to adequately ventilate the home. The fan's cubic-feet-per-minute rating should always be measured at 0.1 inch static pressure and in order to reflect actual air delivery, the manufacturer's recommended louver must be in place.

VENT INSTALLATION STEPS

To install a vent, you will need basic shingling tools, as well as several carpentry tools. Have the following items on hand before you begin work.

- Ladder.
- Roofer's hatchet.
- Tape measures.
- Chalk line and chalk.
- Nail pouch.
- Roofing nails.
- Nail bar.
- Utility knife.
- Roofing cement.
- Trowel.
- Screwdriver.
- Carpenter's level.
- Keyhole saw or saber saw.
- Drill.
- Crayon.
- 3-inch spike.

Turbines   To install wind-driven turbine ventilators, first determine the number of 12-inch ventilators you will need and note their locations. In addition to the turbines, you will need 2 square feet of soffit or eaves-louver area ventilation for every 600 square feet of attic area to allow sufficient outside air to enter the attic.

The best place to install a turbine is on the rear slope of the house about 8 inches from the ridge line and between roof rafters. Drive a 3-inch spike from the inside of the attic through the sheathing and the roof materials. On the roof, use the spike as a center point. With a crayon, lightly mark a circle on the shingles that is wide enough to accommodate the throat of the turbine. Carefully loosen, remove, and save the shingles around the circle. Remove as few shingles as is practical.

Again using the spike as a center point, mark a circle on the felt wide enough to accommodate the throat of the turbine. Remove the spike.

Cut through the sheathing, install the turbine, and replace and trim the shingles. The turbine must be watertight. Refer to the section on installing vent flanges in chapter 7. The techniques for shingling around turbines are nearly identical to shingling around other roof obstacles.

Power Ventilators   Powered ventilators (FIG. 12-11) are installed basically the same way as wind-driven turbines. Depending on the model you select, provisions must be made for thermostat, humidistat, and electrical hookups. Powered ventilators are typically factory-set to automatically switch on at 100 degrees and automatically turn off at 85 degrees.

Ridge Vents   To install ridge vents, first measure along the ridge to determine the number of vents required. Be certain to allow 6 inches at each end of the roof that is not to be cut. In other words, if the ridge measures 35 feet, the total length of ridge vents required will be 34 feet. If a ridge vent must be cut to fit the ridge, make a cut where it will not damage the internal baffle.

Use a utility knife to cut through the cap shingles and felt. Taking care not to cut structural rafters or ridgepoles, use a power saw to make a vent slot to within 6 inches of the end of the ridge. The vent slot should be 2 inches wide for truss construction or $3^1/8$ inches wide for ridgepole construction.

To align the vents, snap chalk lines on both sides of the vent-slot opening in the ridge. Center and install the vents the entire length of the ridge so that the vent ends are flush with both rakes. Nail approximately every 12 inches and at each overlap of joints. Apply caulk as recommended by the manufacturer of the ridge-vent unit you are using.

SKYLIGHTS

When properly sized and oriented, skylights (FIGS. 12-12 and 12-13) can reduce energy consumption throughout the year. Recent improvements in the design of skylights include acrylic-domed units insulated with dead air space, self-flashing models that are fastened directly to the roof deck without a curb, motor-driven vent windows, and models with shades or blinds.

During the summer, electrical lighting costs can be reduced significantly with the use of skylights. Some 90 percent of the electrical energy used for artificial lighting produces heat instead of light. The solar heat gain from skylights will be more than offset by the reduction of heat from decreased use of artificial lighting.

When you are planning the installation of a skylight, keep in mind the type of room in which the skylight is to be installed and how much direct sunlight will enter during each season. A skylight without a suitable interior cover or shade can contribute excessive heat gain. Too much

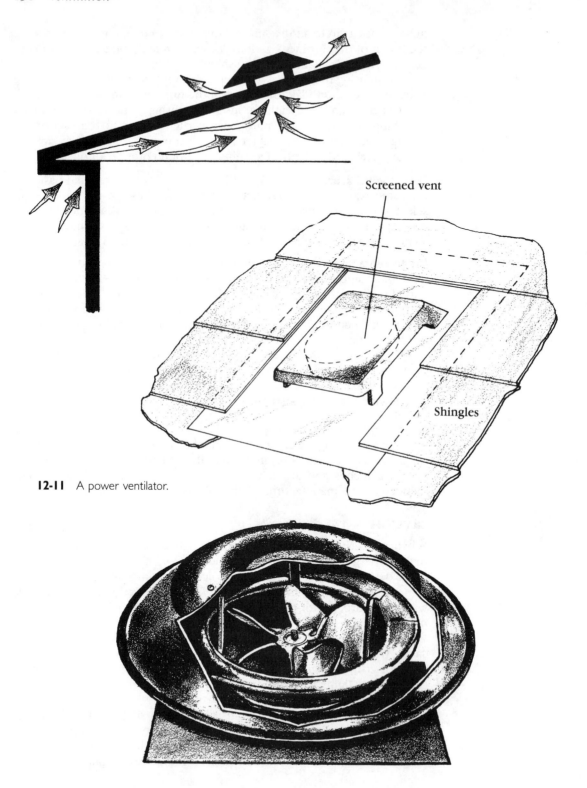

Screened vent

Shingles

12-11 A power ventilator.

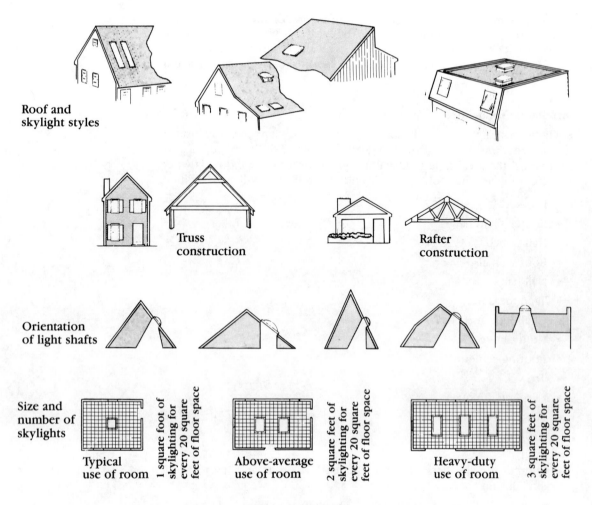

Roof and skylight styles

Truss construction

Rafter construction

Orientation of light shafts

Size and number of skylights

Typical use of room

1 square foot of skylighting for every 20 square feet of floor space

Above-average use of room

2 square feet of skylighting for every 20 square feet of floor space

Heavy-duty use of room

3 square feet of skylighting for every 20 square feet of floor space

12-12 Skylight styles.

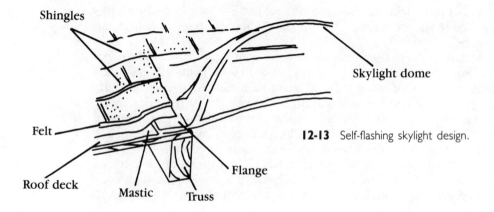

Shingles

Skylight dome

Felt

12-13 Self-flashing skylight design.

Roof deck

Mastic

Truss

Flange

direct sunlight might even bleach carpet, furniture, or walls. One solution is to choose a tinted acrylic skylight. The white, bronze, or gray glazing will help block strong sunlight.

Shafted skylights usually require a mechanical or electrical retracting shade system.

Installation Guidelines

Skylight models vary considerably. Each model will have specific installation instructions with step-by-step details as part of the package. As with roof vents, the ideal time to install a skylight or a roof window is when you are shingling the roof. The tools and materials you will need include:

- Ladder.
- Roofer's hatchet.
- Nail pouch.
- Roofing nails.
- Nail bar.
- Tape measure.
- Chalk line and chalk.
- Utility knife.
- Roofing cement.
- Trowel.
- Carpenter's square.
- Power saw or hand saw.
- Keyhole saw.
- Drill.
- Lumber for headers and for the curb if it is not part of the skylight unit.
- Plasterboard, insulation, etc., if a light shaft is required.

Before you begin work, check the weather forecast and make certain you have allowed sufficient time to complete the project. Depending on the model you select, installation time could take from about 3 hours to 3 days. Before you are ready to position the skylight, arrange for help in carrying the unit to the roof.

Installing a skylight is only a little more complicated than adding roof ventilators. (See FIGS. 12-14 through 12-16.) With a skylight, the size of the opening in the roof is a good deal larger than for vents. The minimum skylight or roof-window area should be 10 percent of the floor space. Another important difference is the additional flashing work required for skylights. With many models, you will have to construct a curb from 2 × 4s. Other units have their own integrated curb and flashing.

The Daylighter LongLite skylight, by APC Corporation, is perhaps the easiest roof window to install because it can be placed directly between existing roof rafters or trusses. A minimum of framing is needed.

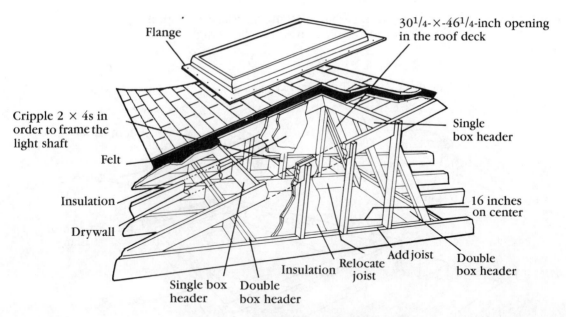

Flange

30¹/4-×-46¹/4-inch opening in the roof deck

Cripple 2 × 4s in order to frame the light shaft

Single box header

Felt

Insulation

16 inches on center

Drywall

Add joist

Double box header

Insulation   Relocate joist

Single box header

Double box header

12-14 Skylight installation details.

The installation position is marked by driving four spikes—one for each corner—from inside the attic through the sheathing and the roof materials. On the roof, use the spikes as guidelines to outline with a crayon the area to be opened. Carefully loosen, remove, and save the shingles around the rectangle. Remove as few shingles as is practical.

Again using the spikes as guidelines, mark a rectangle on the felt. Remove the spikes and cut through the sheathing. At each end of the opening, install a cross brace between the rafters to form a rectangular frame.

Fasten the skylight to the wood framing with clips. A flat flange is part of the unit, and strips of aluminum flashing are nailed where the flange and roof join. Apply roofing cement as directed by the manufacturer as you trim and replace the surrounding shingles.

The Key to Preventing Skylight Leaks

Skylights are flashed and shingled just like any other roof obstacle (see chapter 7). The key to the proper installation of a skylight is the flashing. If a skylight leaks, it will almost always be due to improper installation where the shingles and the flashing join.

At the "top" of the skylight, the first course of shingles must reset snugly against, and be on top of, the aluminum or galvanized sheet-metal collar flashing. Use L-shaped, aluminum or galvanized metal step flashing at the sides of the skylight and weave the shingles with the step flashing.

At the "bottom" of the skylight, the collar flashing must rest on top of one or two courses of shingles (depending on where the skylight meets the shingle courses).

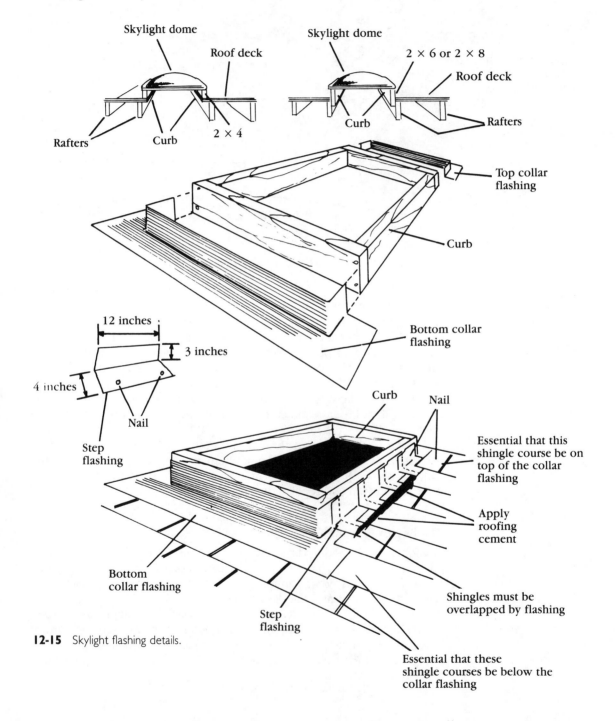

12-15 Skylight flashing details.

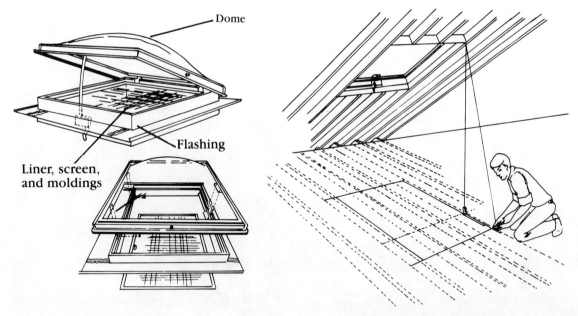

Dome

Flashing

Liner, screen,
and moldings

Carefully lay out the shaft
opening for an angled support system

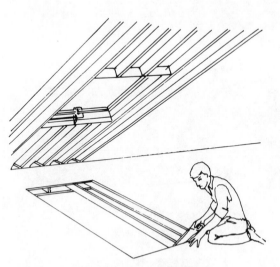

The rough opening in the ceiling
will require insulation on four sides

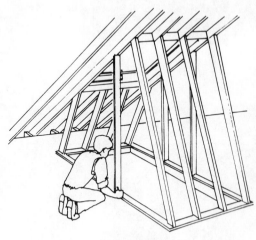

The bottom and top
openings are splayed

12-16 Installing a shafted skylight.

Sources

Associations and Institutes

AMERICAN CANVAS INSTITUTE
10 Beach St.
Berea, OH 44017

AMERICAN WOOD COUNCIL
1619 Massachusetts Ave. NW
Washington, DC 20036

ASPHALT ROOFING MANUFACTURERS ASSOCIATION
1800 Massachusetts Ave. NW
Suite 702
Washington, DC 20036

CALIFORNIA REDWOOD ASSOCIATION
One Lombard St.
San Francisco, CA 94111

HAND TOOLS INSTITUTE
707 Westchester Ave.
White Plains, NY 10604

HOME AIR COMPRESSOR ASSOCIATION
211 East Ontario, Suite 1300
Chicago, IL 60611

HOME VENTILATING INSTITUTE
4300-L Lincoln Ave.
Rolling Meadows, IL 60008

NATIONAL ASSOCIATION OF THE REMODELING INDUSTRY
1901 N. Moore St.
Suite 808
Arlington, VA 22209

NATIONAL PAINT & COATINGS ASSOCIATION
1500 Rhode Island Ave. NW
Washington, DC 20005

NATIONAL ROOFING CONTRACTORS ASSOCIATION
8600 Bryn Mawr Ave.
Chicago, IL 60631
(*Call 800 USA-ROOF for a computerized referral
service for local association members.*)

POWER TOOL INSTITUTE
5105 Tollview Drive
Rolling Meadows, IL 60008

RED CEDAR SHINGLE & HANDSPLIT SHAKE BUREAU
Suite 275
515 116th Ave. NE
Bellevue, WA 98004

SMALL HOMES COUNCIL-BUILDING RESEARCH COUNCIL
University of Illinois
One East St. Mary's Road
Champaign, IL 61820

WESTERN WOOD PRODUCTS ASSOCIATION
1500 Yeon Building
Portland, OR 97204

Asphalt- and Fiberglass-Based Shingles and Roll Roofing

BIRD & SON
East Walpole, MA 02032

CELOTEX CORPORATION
Roofing Products Division
P.O. Box 22602
Tampa, FL 33622

CERTAINTEED CORPORATION
P.O. Box 860
Valley Forge, PA 19482

CHAMPION INTERNATIONAL CORPORATION
Building Products Division
Stamford, CT 06921

ELK CORPORATION
P.O. Box 2450
Tuscaloosa, AL 35403

FLINTKOTE ROOFING PRODUCTS
580 Decker Dr.
Irving, TX 75062

GAF CORPORATION
140 West 51 St.
New York, NY 10020

GEORGIA-PACIFIC
133 Peachtree St. NE
Atlanta, GA 30348-5605

W. R. GRACE & COMPANY
Construction Products Division
Ice & Water Shield Membrane
62 Whittlemore Ave.
Cambridge, MA 02140-1623

JOHNS-MANVILLE
Ken-Caryl Ranch
Denver, CO 80217

ONDULINE ROOFING PRODUCTS
4900 Onduline Drive
Fredricksburg, VA 22401

OWENS-CORNING ROOFING PRODUCTS
997 Old Eagle School Rd.
Suite 215
Wayne, PA 19087

TARMAC ROOFING SYSTEMS INC.
E-Z Roll Roofing
1401 Silverside Road
Wilmington, DE 19810

Caulks and Sealants

DAP INCORPORATED
P.O. Box 277
Dayton, OH 44501

GIBSON-HOMANS COMPANY
1755 Enterprise Parkway
Twinsburg, OH 44087

MACCO ADHESIVES
30400 Lakeland Blvd.
Wickliffe, OH 44092

RED DEVIL INC.
2400 Vauxhall Rd.
Union, NJ 07083

3M
P.O. Box 4039
St. Paul, MN 55104

Cedar Shingles and Shakes

KOPPERS COMPANY
437 Seventh Ave.
Pittsburgh, PA 15219

MASONITE CORPORATION
329 N. Wacker Dr.
Chicago, IL 60606

SHAKERTOWN ROOFING AND SIDING
P.O. Box 400
Winlock, WA 98596

Fasteners

ARROW FASTENER COMPANY, INC.
271 Maybill St.
Saddle Brook, NJ 07662

BOSTITCH DIVISION
Textron Inc.
Briggs Drive
East Greenwich, RI 02818

DUO-FAST CORPORATION
3702 N. River Rd.
Franklin Park, IL 60131

KEYSTONE STEEL & WIRE COMPANY
Red Brand Nails
7000 South West Adams
Peoria, IL 61641

PHIFER WIRE PRODUCTS, INC.
Aluminum Nails
P.O. Box 1700
Tuscaloosa, AL 35403

SENCO FASTENERS
8485 Broadwell Rd.
Cincinnati, OH 45244

SWINGLINE FASTENERS
32-00 Skillman Ave.
Long Island City, NY 11101

TECO WOOD FASTENERS
5530 Wisconsin Ave.
Washington, DC 20015

VACO TOOLS & FASTENERS
1510 Skoki Blvd.
Northbrook, IL 60062

Hand Tools

AMERICAN TOOL COMPANIES, INC.
Prosnip Aviation Snips
P.O. Box 337
DeWitt, NB 68341

AMES CONTRACTOR CATALOG
505 O'Neill Ave.
Belmont, CA 94002

ARDELL INDUSTRIES, INC.
P.O. Box 1573
555 Lehigh Ave.
Union, NJ 07083

MCGUIRE-NICHOLAS MFG. CO.
6223 Santa Monica Blvd.
Los Angeles, CA 90038

THE STANLEY WORKS
P.O. Box 1800
New Britain, CT 06050

U.S. GENERAL
Tool Catalog
100 Commercial St.
Plainview, NY 11803

VAUGHAN & BUSHNELL MFG. CO.
11414 Maple Ave.
Hebron, IL 60034

Ladders and Scaffolds

KELLER LADDERS
18000 State Rd. Nine
Miami, FL 33162

R.D. WERNER CO., INC.
Greenville, PA 16125

Metal Roofing

ALCAN BUILDING PRODUCTS
2401 Parkman Rd. NW
Warren, OH 44485

ALCOA BUILDING PRODUCTS, INC.
Suite 1200
Two Allegheny Center
Pittsburgh, PA 15212

FOLLANSBEE STEEL CORPORATION
Follansbee, WV 26037

REINKE SHAKES
210 South 4th St.
P.O. Drawer 88
Hebron, NB 68370

REYNOLDS METALS COMPANY
Architectural & Building Products
P.O. Box 25670
Richmond, VA 23261

UNITED STATES STEEL CORPORATION
600 Grant St.
Pittsburgh, PA 15230

WHEELING-PITTSBURGH STEEL CORPORATION
Wheeling Corrugating Co.
1134-40 Market St.
Wheeling, WV 26003

Publications and Videocassettes

CONSUMER INFORMATION CATALOG
Pueblo, CO 81009

"HOMETIME" HOME IMPROVEMENT VIDEOCASSETTES
6213 Bury Drive
Eden Prairie, MN 55346
Phone 1-800-345-8000

SUPERINTENDENT OF DOCUMENTS
U.S. Government Printing Office
Washington, DC 20402

U.S. DEPARTMENT OF COMMERCE
National Technical Information Service
Springfield, VA 22161

Roof Decking and Insulation

AKOMA CORPORATION
Thermaseal
1570 Halgren Road
Maple Plains, MN 55359

ARCO CHEMICAL COMPANY
1500 Market Street
Philadelphia, PA 19101

CELOTEX CORPORATION
Thermax Sheathing
1500 N. Dale Mabry
Tampa, FL 33607

DOW CHEMICAL
Styrofoam
2020 Dow Center
Midland, MI 48640

HOMASOTE COMPANY
P.O. Box 7240
West Trenton, NJ 08628

RMAX, Inc.
Therma Sheath
13524 Welch Road
Dallas, TX 75234

U.S. GYPSUM COMPANY
101 S. Wacker Drive
Chicago, IL 60606

Skylights

ANDERSEN CORPORATION
Box 12
Bayport, MN 55003

APC CORPORATION
50 Utter Ave.
P.O. Box 515
Hawthorne, NJ 07507

FOX MARKETING INC.
4518 Taylorsville Rd.
Dayton, OH 45424

NATURALITE, INC.
P.O. Box 28636
Dallas, TX 75228

ODL INCORPORATED
215 E. Roosevelt Ave.
Zeeland, MI 49464

ONDULINE ROOFING PRODUCTS
4900 Onduline Drive
Fredericksburg, VA 22401

PELLA SKYLIGHTS
Rolscreen Company
Pella, IA 50219

PLASTECO, INC.
P.O. Box 24158
Houston, TX 77229

PPG INDUSTRIES, INC.
P.O. Box 16012
Pittsburgh, PA 15242

ROLLAMATIC ROOFS INC.
1400 Yosemite Ave.
San Francisco, CA 94124

SKYMASTER SKYLIGHTS
413 Virginia Dr.
Orlando, FL 32803

VELUX-AMERICA INC.
74 Cummings Park
Woburn, MA 01801

VENTARAMA SKYLIGHT CORPORATION
140 Cantiague Rock Rd.
Hicksville, NY 11801

WASCO PRODUCTS, INC.
Pioneer Ave.
P.O. Box 351
Stanford, ME 04073

Ventilators

AIR VENT INC.
4801 N. Prospect Rd
Peoria Heights, IL 61614

ALCOA BUILDING PRODUCTS
P.O. Box 716
Sidney, OH 45365

ARVIN INDUSTRIES, INC.
Arvin Wind Turbines
Columbus, IN 47201

AUBREY MANUFACTURING, INC.
6709 South Main St.
Union, IL 60180

LEIGH PRODUCTS
Coopersville, MI 49404

LOMANCO
P.O. Box 519
Jacksonville, AK 72076

MIDGET LOUVER COMPANY
800 Main Ave.
Route 7
Norwalk, CT 06852

NAUTILUS
Roof-Mount Powered Attic Ventilators
Hartford, WI 53027

ROBBINS & MYERS,INC.
Hunter Ventilating and Circulating Fans
2500 Frisco Ave.
Memphis, TN 38114

STEPHENSON CUPOLA
Conneaut, OH 44030

Glossary

abut To position snugly against the top or side of a shingle. Shingle courses can be abutted when a new layer is applied over worn, but not warped or buckled, shingles; butt.

alignment notch Factory-cut end of a shingle where cutouts meet to form proper shingle alignment.

asphalt fiberglass-based shingles Shingles made with an inorganic base saturated with asphalt and topped by colored ceramic granules or opaque rock for weather and sunlight resistance.

attic ventilator Screened openings located in the soffit, gable ends, or at the ridge line. Power-driven fans can be used as part of a balanced exhaust system for homes.

aviation snips Tin snips can be used to trim asphalt-fiberglass shingles as well as metal flashing.

backing down High nailing courses of roofing material to tie in lower, successive courses.

backing in *See* filling in.

blind nailing Installing nails so that the nail heads are concealed by roofing material.

border shingles Roofing material applied to the outer edges of a roof section to provide protection and to present an even edge.

building paper Light, asphalt-saturated material used to temporarily waterproof a roof deck; felt.

build-up roofing Applied to flat roofs with three to five layers of asphalt-saturated felt and roll roofing. Designed to hold water until it evaporates rather than shed water, hot tar is mopped and finished with crushed slag or gravel. Not appropriate for installation by do-it-yourselfers. Special equipment is required.

butt *See* abut.

capping   One-tab shingles centered and applied horizontally to the ridge or hip of a roof; the last course of roll roofing centered and applied horizontally to the ridge.

caulk   Waterproofing material applied to chimney flashing seams or vent-pipe seams. Not a replacement for roofing cement.

ceramic granules   Fire-hardened material added to asphalt and fiberglass shingles to provide color and for weather resistance.

claw hammer   A carpenter's hammer can be used to nail shingles, but the lightweight head is a disadvantage for roofers.

closed valley   Where two internally sloping sections intersect, shingles are weaved instead of cut to form a channel.

collar   *See* flange.

counter flashing   Metal strips used to prevent moisture from entering the top edge of roof flashing, as on a chimney or wall.

course   One series of shingles in a horizontal, vertical, or 45-degree angle from the eaves to the ridge. Shingles, shakes, and roll roofing are laid in successive courses.

coverage   Amount of weather protection provided by shingles or other roofing material. Properly installed shingles provide double coverage and usually are installed with 5 or $5^1/8$ inches exposed to the weather.

cricket   Small, sloped structure made of metal and designed to drain moisture away from a chimney. Usually placed at the back of a chimney.

cupola   Square or round roof structure sometimes capped by a weather vane. Often used to enhance attic ventilation.

cutout   Where alignment notches of three-tab shingles meet to create a pattern; the watermark.

d   Abbreviation indicating penny sizing per 100 of some nails.

deck   Materials such as plywood sheathing or planking installed over framing members. The roof surface before shingles or other roofing materials are installed.

drip edge   Lightweight metal strips designed to fit snugly against rakes and eaves. They are chiefly cosmetic but will provide some protection for the deck edges.

eaves   The bottom edge of the roof deck projecting over the wall of a building.

exposure   Portion of the shingle subjected to the weather, usually 5 or $5^1/8$ inches from the butt of one shingle to another with three-tab shingles.

felt   Light, asphalt-saturated building paper used to temporarily waterproof a roof deck; an underlayment for the smooth application of shingles and roll roofing.

filling in   Roofing a section by squaring off an angled portion to obtain long vertical runs of shingles.

flange   Metal or plastic trim fitted over pipe or other venting unit to waterproof the intrusion in the roof deck.

flashing   Strips of copper, aluminum, or galvanized sheet metal used along walls, dormers, valleys, and chimneys to prevent moisture seepage.

45 pattern   With three-tab shingles, application resembling steps at a 45-degree angle. Shingle cutouts align at every other course.

furring strips   Lightweight wood strips applied as supplemental fastening to temporarily prevent wind damage to felt on a roof deck; fastening base for wood roofing materials.

gable   Triangular end portion of a building from the eaves to the ridge of the roof.

gauge   Diameter of the wire used to manufacture nails. The adjustable device on a roofer's hatchet used to align proper shingle exposure.

granule   Finely crushed or ground minerals, sand, or rock adhered to the portion of shingles and roll roofing exposed to the weather to provide color and weather resistance.

hatchet   Designed for professional roofers, shingling hatchets have a large, square head with a milled face. An adjustable gauge permits proper shingle exposure and alignment.

high nailing   Driving nails about 1 inch below the top of the shingle (which is well above the standard nailing location) to properly positon one course of shingles while allowing one or more courses to be added underneath the high-nailed course. Each shingle eventually must be properly nailed.

hip pad   A rubber pad strapped to the shingler's hip to protect clothing from excessive wear.

hip roof   Angle formed by the meeting of sloping roof sections. Slopes are angled toward the center from four sides and there are no rakes.

laddeveyor   Combination ladder and conveyor belt designed to lift bundles to roofs.

louver   Ventilating unit with slats; usually placed at gables.

mansard roof   Steeply sloped portion of a two-part roof deck, often having one or more windows, where shingles generally must be applied from a ladder or a scaffolding system.

mastic   Roof coating used for thermal insulation or waterproofing around vents and other roof obstacles.

nail bar   Tool for removing nails and flashing.

open valley   Where two internally sloping roof sections intersect, shingles are trimmed to form a channel.

peak   *See* ridge.

penny   Measure of nail length indicated by the letter d; originally an indicator of price per 100 nails.

pitch   The incline, or slope, of a roof; the ratio of the total rise to the total width of a house measured in inches per rise per foot of run.

purlins   Horizontal wood support strips between the roof frame's plate and the ridge.

rake   Edges of the roof deck running parallel to the slope.

random-spacing pattern   With three-tab shingles, a pattern with cutouts aligned every six shingles.

ridge   The horizontal junction of the two top edges of two sloping roof sections.

ripping hammer   A carpenter's hammer can be used to install shingles but, like the claw hammer, the lightweight head is not efficient for roofers.

roll roofing   Asphalt-based roofing material weighing 45 to 90 pounds per roll and laid horizontally over low-pitched roofs.

roof jack   Metal device used to support scaffolding on steeply pitched roofs.

run   The distance covered by the application of shingles in one pass of the pattern; an inclined course.

seal-down strip   Factory-applied, sunlight-activated adhesive that bonds asphalt-based fiberglass shingles to the course above.

shake   Hand-split, edge-grained wood shingle.

sheathing   Plywood or planks that form the roof deck.

shingle   Roof covering made from asphalt, fiberglass, wood, aluminum, tile, slate, or other water-shedding material.

slope   Degree of incline of a roof plane usually given in inches of rise per horizontal foot of the run.

soffit   Underside of the eaves.

square   When used as a reference to roofing material, 100 square feet of coverage.

starter course   First row of shingles or roll roofing applied at the eaves.

step flashing   Usually aluminum or galvanized sheet metal cut in L-shaped pieces that are weaved at the joints between the roof surface and the roofing material; installed along walls and masonry.

straight pattern   Vertical application of three-tab shingles with cutouts aligned every other shingle.

tab   On asphalt and fiberglass shingles, the material between the factory cutouts.

tin snips   *See* aviation snips.

valley   Internal angle formed by the junction of two sloping roof sections.

vent collar   *See* flange.

watermark   Line where double coverage of roofing material laps courses. *See* cutout.

whole-house fan   Ventilator usually installed in an attic floor to exhaust air from the entire house.

wing   Roof section broadly extended or projecting at an angle from the main building.

Index

roof line louvers, 191
shingle removal around, 39
shingling around, 136-139
sources and information for,
  212
turbine, 188, 191, 194, 196
under-eaves, 191, 193
whole-house fan and, 188,
  192, 195
videocassettes, 210

W

walls
  shingling up to, 140-144
  step flashing for, 140-142
  trimming shingles along,
    142-144

weather factors, safety and, 23
whole-house fan, ventilators
  and, 188, 192, 195
winged roof, 101-103
wood shingles and shakes, 3, 6,
  149-161
  alignment of, 151
  application steps for, 152-
    161
  capping ridge with, 160
  chimneys and flashing for,
    156
  deck or furring strips for, 149
  estimating amount needed,
    152
  exposure recommendations
    for, 153

flashing for, 149
hip roofs, 155, 158
metal look-alikes, 163-166
mitering joints for, 155, 158
nailing techniques for, 155
nails for, 152
panelized roofing, 160-161
pipes and pipe collars, 159
sources and information for,
  208
spacing for, 151, 154
stains and coatings for, 152
starter course for, 153, 157
tools and equipment for, 150
valleys, 156, 158
ventilation strips for, 153,
  154